Plant-Based Bioactive Substances Identification, Extraction, and Application

Plant-Based Bioactive Substances Identification, Extraction, and Application

Guest Editors

Elson S. Alvarenga
Wei Li

MDPI

Basel • Beijing • Wuhan • Barcelona • Belgrade • Novi Sad • Cluj • Manchester

Guest Editors

Elson S. Alvarenga
Department of Chemistry
Universidade Federal de Viçosa
Viçosa
Brazil

Wei Li
Department of Pharmacognosy
Toho University
Funabashi
Japan

Editorial Office
MDPI AG
Grosspeteranlage 5
4052 Basel, Switzerland

This is a reprint of the Special Issue, published open access by the journal *Plants* (ISSN 2223-7747), freely accessible at: https://www.mdpi.com/journal/plants/special_issues/Plant_Bioactive.

For citation purposes, cite each article independently as indicated on the article page online and as indicated below:

Lastname, A.A.; Lastname, B.B. Article Title. *Journal Name* **Year**, *Volume Number*, Page Range.

ISBN 978-3-7258-4689-4 (Hbk)
ISBN 978-3-7258-4690-0 (PDF)
https://doi.org/10.3390/books978-3-7258-4690-0

Contents

About the Editors

Elson S. Alvarenga

Elson S. Alvarenga is a Full Professor at the Universidade Federal de Viçosa (UFV), located in Viçosa, Brazil. His research is primarily dedicated to the synthesis of analogues of natural products with potential biological activity, particularly in the context of sustainable agriculture. To fully characterize these synthesized compounds, his work integrates a range of spectrometric techniques—such as NMR, MS, and IR—alongside theoretical calculations to elucidate their structural and electronic properties. He also employs molecular docking studies to identify possible enzymatic targets of these bioactive compounds, providing insights into their mechanisms of action. Although his research spans several areas, a key emphasis is placed on the bioanalytical evaluation of insecticidal and herbicidal activities. His overarching objective is to discover new bioactive molecules that are effective against weeds and insect pests while posing minimal risk to crops, beneficial insects (such as pollinators), and natural enemies within ecosystems. By aligning synthetic chemistry with biological testing and computational modelling, Professor Alvarenga aims to contribute to the development of environmentally safer agrochemicals that support both productivity and ecological balance.

Wei Li

Wei Li heads the Department of Pharmacognosy, Faculty of Pharmaceutical Sciences, Toho University. He received a PhD in Pharmacognosy from Kitasato University Graduate School of Pharmaceutical Sciences in 2000. After that, he joined Toho University, and he has been engaged in the study of natural medicine chemistry since. His main research interests focus on the discovery of novel bioactive natural products from traditional medicines. He has published over 200 scientific papers. At present, he serves as a member of several editorial boards across journals such as the *Journal of Natural Medicines*. He has also received several awards, including "The JSP Award for Scientific Contributions" from the Japanese Society of Pharmacognosy.

Preface

It is with great pleasure that we present this Reprint of "Plant-Based Bioactive Substances Identification, Extraction, and Application", a Special Issue of *Plants* (MDPI) which brings together nine scholarly works.

This compilation highlights recent advances in the exploration and application of plant-derived natural products with biological activity, reflecting the growing interest in the use of sustainable alternatives in agriculture, medicine, and environmental management. Given the well-established bioactivity of allelochemicals—natural compounds that influence the growth and development of other organisms—and the diverse pharmacological potential of natural epoxides, one study focuses on synthesizing four novel epoxy sesquiterpene lactones and the related derivatives based on dehydrocostus lactone, a sesquiterpene known for its broad spectrum of biological effects. The herbicidal activities of these compounds were evaluated, with the goal of identifying effective and environmentally friendly agents for weed management.

Another investigation assessed the anticancer potential of *Cymbopogon citratus* (lemongrass) essential oil against several human cancer cell lines (A375, RPMI-7951, MCF-7, and HT-29), conducting a comparative analysis with citral, its main component, as well as exploring the essential oil's molecular mechanisms of action using in silico modeling tools.

In the realm of agrochemical formulation, binary mixtures of bioactive compounds with herbicidal or biopesticidal properties were developed using emulsifiers to stabilize their volatile constituents. Aiming to develop multi-target natural herbicides with minimal environmental impact, these formulations were tested for their physicochemical stability and herbicidal activity, including their impact on Photosystem II, wheat germination, and seedling development.

Aquatic plant–microcrustacean interactions were also explored, with one contribution studying the effects of three submerged macrophytes (*Ceratophyllum demersum*, *Myriophyllum aquaticum*, and *Stuckenia pectinata*) and their exudates on the population dynamics, feeding, and reproduction of *Daphnia* cf. *pulex* and *Simocephalus* cf. *mixtus*, providing essential insights for understanding aquatic ecosystem functioning.

The influence of developmental stages and environmental conditions on *Gladiolus* resistance to western flower thrips (WFTs) was evaluated through metabolomic profiling across three growth stages in two cultivars—Robinetta (resistant) and Charming Beauty (susceptible). The study's findings underscore the importance of developmental timing in integrated pest management strategies.

The pharmacological exploration of *Wikstroemia indica*, a medicinal plant from southeastern China, led to the identification of daphnane diterpenoids—compounds previously undocumented in this species—using LC-MS/MS, opening new avenues for drug discovery.

Finally, researchers quantified the content of argentatins and isoargentatins A and B in the leaves and stems of three *Parthenium argentatum* (guayule) accessions under different irrigation conditions, contributing to ongoing efforts to optimize phytochemical yields in industrial crops.

Elson S. Alvarenga and Wei Li
Guest Editors

plants

Article

Exploring Sesquiterpene Lactones from *Saussurea lappa*: Isolation, Structural Modifications, and Herbicide Bioassay Evaluation

Elson S. Alvarenga [1,*], Francisco A. Macías [2], Stephani S. Ferreira [1], Juan C. G. Galindo [2] and José M. G. Molinillo [2]

[1] Department of Chemistry, Universidade Federal de Viçosa, Viçosa 36570-900, MG, Brazil; stephani.ferreira@ufv.br
[2] Department of Organic Chemistry, Faculty of Sciences, Universidad de Cádiz, c/República Saharui s/n, Puerto Real, 11510 Cádiz, Spain; famacias@uca.es (F.A.M.); juancarlos.galindo@uca.es (J.C.G.G.); chema.gonzalez@uca.es (J.M.G.M.)
* Correspondence: elson@ufv.br

Abstract: Considering the resistance of weeds to different herbicides with different mechanisms of action, the search for new, more selective compounds with low toxicity to other species in nature has been very important for the development of agriculture. Because of that, considering the biological activity of allelochemicals and natural epoxides, four new epoxy compounds derived from dehydrocostus lactone were synthetized and evaluated for their potential herbicide activity against three species of seeds, *Allium cepa* (onion), *Lepidium sativum* (garden cress), and *Lactuca sativa* (lettuce). In assays with *A. cepa*, compound **4** inhibited radicle length by 80% at 100 µM. Notably, for *L. sativum*, compound **4** showed significant inhibition, reducing stalk and radicle lengths by 80% at 100 µM, surpassing the performance of the commercial herbicide Logran. However, diol **5** notably inhibited radicle growth by 28% at 100 µM, making the most significant observed effect. One of the noteworthy lactones studied is epoxide **4**. This highlights the importance of the epoxide functional group in affecting both radicle and shoot lengths of seeds. Therefore, the synthesis of these compounds has proven advantageous and holds great potential for the development of new herbicides.

Keywords: sesquiterpene lactone; herbicide; coleoptile; dehydrocostus lactone; DHC; epoxide

1. Introduction

The constant search for efficient herbicides is of the utmost importance for the agricultural industry. This is due to the presence of pests, which substantially affect global agricultural production, negatively interfering with the productivity of the crop, the final quality of the products, and consequently the production costs, mainly because of the relationship of competition of the culture with the plants for water, light, and nutrients [1,2].

Since the 20th century, synthetic herbicides have been used to control these unwanted plants; however, the constant and excessive use of herbs with the same mechanism of action (MoA) in large areas of cultivation, has resulted in the slow evolution and resistance of these pests against the exploited agrochemicals [1]. Thus, there is a growing demand for new, efficient herbicides with alternative mechanisms of action and low environmental impact [3,4].

In this context, natural products can be considered a viable option to replace traditional synthetic agrochemicals [5,6]. This is because they can be synthesized by living organisms, as well as produced through complex synthetic pathways that lead to molecules employing mechanisms of action different from those used by current agrochemicals [7]. Furthermore, herbicides like phosphinothricin and phosalacin, derived from natural products, underscore the feasibility of studying and evaluating these compounds as potential future herbicides [8].

The sesquiterpenic lactones, one of the largest families of natural products, are compounds widely distributed in plants and more than 7000 structures of this class have already been described [9,10]. In addition, studies referring to these compounds indicate that they have a broad biological potential, standing out in terms of cytotoxic, antitumoral, antibacterial, anti-inflammatory, antimalaric, and antifungal activities [11–14].

These natural compounds can be isolated mainly from the aerial parts of certain plants of the *Asteraceae* family, also being found in families such as *Umbelliferae, Lauraceae,* or *Magnoliaceae* [7]. Sesquiterpenes also represent an important group of secondary metabolites of the *Asteraceae* family [15–17].

Sesquiterpenic lactones also have great potential for application in the agricultural sector [16,18]. Studies show that some secondary metabolites can affect the germination and growth of certain plant species, but the nature and the extent of the effects produced depend on several factors, such as the structure of the lactone tested, its concentration, and the species on which it will act [19–21].

This effect, called allelopathy, is of great importance in agriculture for pest control, provided that a correct association is established between the species producing allelopatic substances and the cultivated species [22,23]. Sesquiterpene lactones such as estafiatin, arglabin, arborescin, and heliangine serve as examples of compounds with allelopathic effects, producing chemicals that negatively impact the growth and development of competing plants [20] (Figure 1).

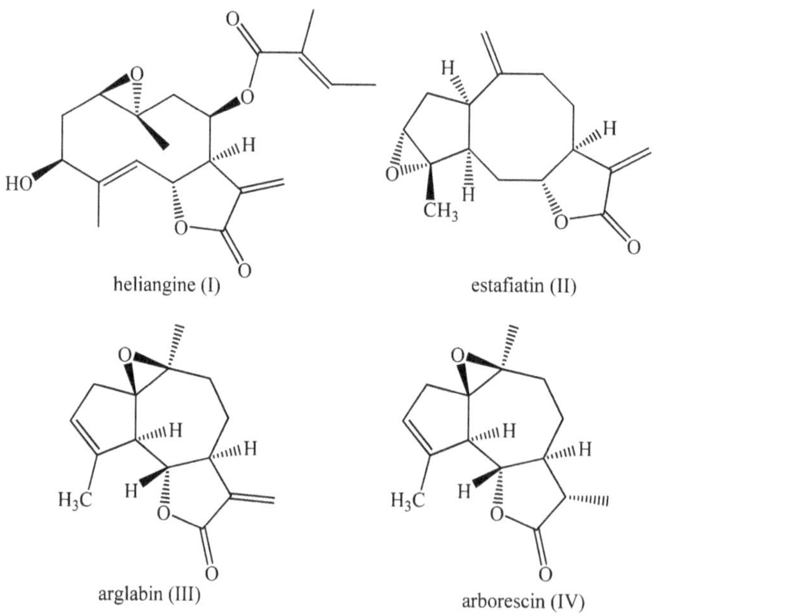

Figure 1. Sesquiterpene lactones: heliangine (**I**), estafiatin (**II**), arglabin (**III**), and arborescin (**IV**).

The compounds arglabin and arborescin, guaianolides structurally related to estafiatin [24], are isolated natural epoxides of *Artemisia glabella* and *Artemissa arborescens*, respectively [25,26]. The stereoselective syntheses of these natural products are described in the literature [25,26]. As for the compound heliangine, it is an epoxide of a germacranolide with plant physiological activity, isolated from *Helinthus tuberosus* L., which can act as a regulator of plant growth [27,28].

Studies suggest that the α-methylene-butirolactone group, present in many natural sesquiterpene lactones (SLs), plays an important role in the bioactivity of these compounds [29,30]. This group has been proposed as one of the structural aspects that can determine their allelopathic activity, as well as their biological activity in general [29–31].

The activity of SLs is linked to their α-methylene-butirolactone component, which serves as a potent and selective alkylating agent for nucleophilic substrates. In fact, the primary determinant of SL cytotoxicity is the presence of an α,β-unsaturated keto group, which is not necessarily associated with the lactone's carbonyl group [7,32]. Additionally, another study found that the absence of this system does not significantly diminish the growth of weeds [7]. The spatial arrangement of the carbon skeleton has been linked to the lower activity of sesquiterpene lactones. The eudesmanolides reynosin and santamarin feature a backbone that closely resembles the spatial arrangement found in germacranolides and also possess an α,β-unsaturated carbonyl system. Compounds with a "double crown"-like spatial configuration (such as germacranolides and eudesmanolides) resemble strigolactones and can fit into the receptor's cavity. However, the observed activity of compounds lacking the unsaturated double bond in the lactone ring remains unexplained. Further research exploring precise biochemical pathways is needed, reinforcing its potential as a sustainable resource for enhancing herbicidal activity and well-being [33–36].

Thus, given the well-documented biological activity of allelochemicals—natural compounds that influence the growth and development of other organisms—and natural epoxides, which are known for their diverse bioactive properties, our research is directed toward the synthesis of four novel epoxy sesquiterpene lactones and other derivatives. These derivatives are designed from dehydrocostus lactone, a sesquiterpene lactone that has already been recognized for exhibiting multiple biological activities [37,38]. The primary objective of this study is to evaluate the potential herbicidal activity of these newly synthesized compounds, assessing their effectiveness in inhibiting plant growth and exploring their possible application as natural herbicides in agricultural weed management.

2. Results and Discussion

2.1. Synthesis

For the study of the herbicidal activity of sesquiterpene lactones, new epoxy derivatives were synthesized from dehydrocostus lactone (DHC, **V**), yielding derivatives (**1**), (**2**), (**3**), (**4**), and (**5**). These compounds were obtained from the reaction of DHC with *meta*-chloroperbenzoic acid (MCPBA) in anhydrous DCM, with yields of 3%, 64%, 16%, 21%, and 6%, respectively, as shown in Figure 2.

As expected, compounds **1–3** exhibited similar spectrometric data, as they differ only in the positions and relative stereochemistry of the epoxide moieties and hydroxyl groups attached to the new components. The IR spectra display peaks in the range of 1765–1770 cm^{-1}, indicating the presence of lactone carbonyl groups, and a band near 1260 cm^{-1}, attributed to C–O stretching. Additionally, compound **5** shows a peak at 3472 cm^{-1}, indicating the presence of a hydroxyl group [39].

The ^{1}H NMR and ^{13}C NMR spectra were crucial for the identification of the compounds, providing detailed information about their molecular structure, including the chemical environment of hydrogen and carbon atoms. These spectra allowed for the accurate determination of functional groups, the number of protons and carbons, as well as their connectivity within the molecule.

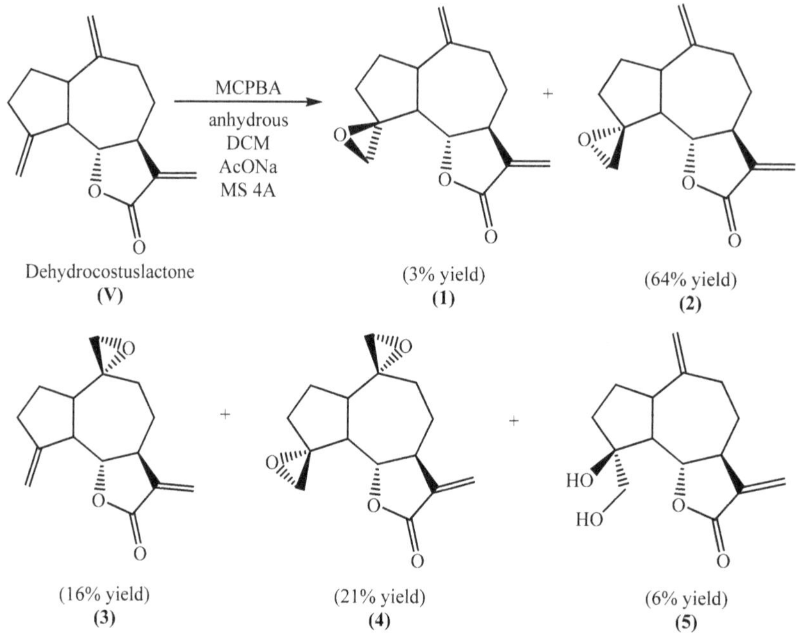

Figure 2. Epoxidation reaction of dehydrocostus lactone (**V**).

In the ^{13}C NMR spectra, signals between $\delta = 168.0$ and 169.5 ppm observed in all five spectra correspond to a carbonyl group. Signals from sp^2-hybridized carbon and carbon bonded to electronegative atoms, such as those in epoxides and hydroxyl-bonded carbons in molecule **5**, are exhibited in the deshielded regions of the spectrum.

In the ^1H NMR spectra, the signals corresponding to hydrogens bonded to sp^2 carbons are found in the deshielded regions of the spectrum due to the anisotropic effect occurring in double bonds. An example is the signal of hydrogen H13 at $\delta = 6.06$–6.07 ppm in molecule 1, appearing as a doublet of doublets (J = 0.4, 2.8 Hz), as shown in Figure S9 in the Supplementary Information (SI). This technique also aids in the analysis of hydrogen bonded to sp^3 carbons, resulting in signals in more shielded regions of the spectrum, such as the signal of hydrogen H8 at $\delta = 1.29$–1.36 ppm in molecule 2. This signal, integrated for one hydrogen, appears as a multiplet in Figure S13 (SI).

The analyses of COSY (correlation spectroscopy), HMBC (heteronuclear multiple bond coherence), and HMQC (heteronuclear multiple quantum correlation) were also equally important for the complete assignment of the signals [40].

To determine the relative configuration of the molecules by NMR spectroscopy, the spatial distances between the hydrogen atoms of the molecules were measured using Overhauser nuclear effect (NOE) experiments. These experiments provide crucial information about through-space interactions between nuclei, allowing for the identification of proximity relationships within the molecular structure. By analyzing NOE enhancements, it is possible to infer the three-dimensional arrangement of atoms, which aids in distinguishing between different stereochemical configurations. This approach is particularly useful for rigid molecules in solution, where direct bond connectivity alone may not be sufficient to establish the full spatial arrangement [41].

As previously described, compound 5 was also utilized in the synthesis of 5α-hydroxy-isozaluzanine C (**6**) through a dihydroxylation reaction. This transformation was carried out using DHC as the starting material, with selenium oxide and *tert*-butyl hydroperoxide serving as the key oxidizing agents. The reaction conditions were carefully controlled to promote selective oxidation, leading to the formation of the hydroxylated product. This method provides an efficient approach to modifying the molecular structure, potentially enhancing its biological activity. The reaction pathway and structural changes involved in this transformation are illustrated in Figure 3.

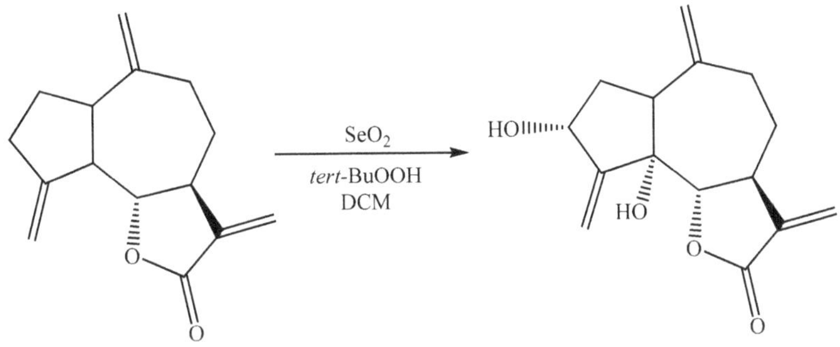

DHC (**V**) (**6**)

Figure 3. Dihydroxilation of dehydrocostuslactone.

The formation of 13-oxo-epi-costuslactone (**7**) involved a two-step synthetic process starting with the reaction of DHC with aqueous sodium carbonate in hexamethylphosphoramide (HMPA). Following this transformation, the primary alcohol underwent oxidation using Dess–Martin periodinane, a mild and selective oxidizing agent commonly employed for converting primary alcohols to aldehydes. This oxidation step was crucial for achieving the desired structural modification, ultimately leading to the formation of 13-oxo-epi-costuslactone (Figure 4).

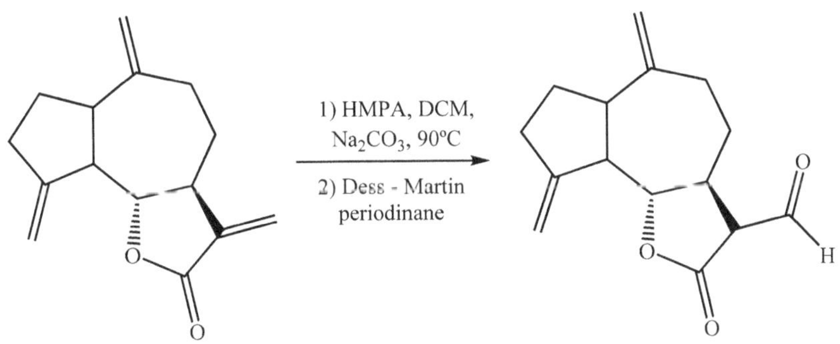

DHC (**V**) (**7**)

Figure 4. Michael addition of a nucleophilic hydroxyl group to DHC and Dess–Martin periodinane oxidation.

2.2. Coleoptile Bioactivity

The study of coleoptile bioactivity provides valuable information about the mechanisms of action of bioactive substances and their potential agricultural applications. The wheat coleoptile bioassay is used to evaluate plant growth stimulation and assess the potential of natural products, synthetic compounds, or agricultural chemicals as herbicides.

Tests using etiolated wheat coleoptiles have been employed to assess the herbicidal potential of compounds prepared by microwave irradiation of costunolide [42], extracts from the leaves of *Origanum majorana* L. [43], steroidal saponins [44], furanocoumarins isolated from the aerial parts of *Ducrosia anethifolia* [45], and many other sources.

Seven compounds were obtained from DHC and subjected to a bioassay using etiolated wheat coleoptiles. Six dilutions (1000, 600, 300, 100, 30, and 10 μmol L^{-1}) were used in the assay. The etiolated wheat coleoptile bioassay was chosen as a preliminary method to evaluate the bioactivity of compounds **1–7** due to its sensitivity to a broad spectrum of bioactive substances.

The results shown in Figure 5 indicate that all compounds exhibited high inhibitory activity at a concentration of 1000 μM, with compounds **1** and **2** standing out. A decrease in inhibitory activity was observed as the concentration was reduced. Notably, even at lower concentrations, compounds **1** and **2** remained the most active, confirming that compounds **1** and **2** were the most effective in the coleoptile bioassay.

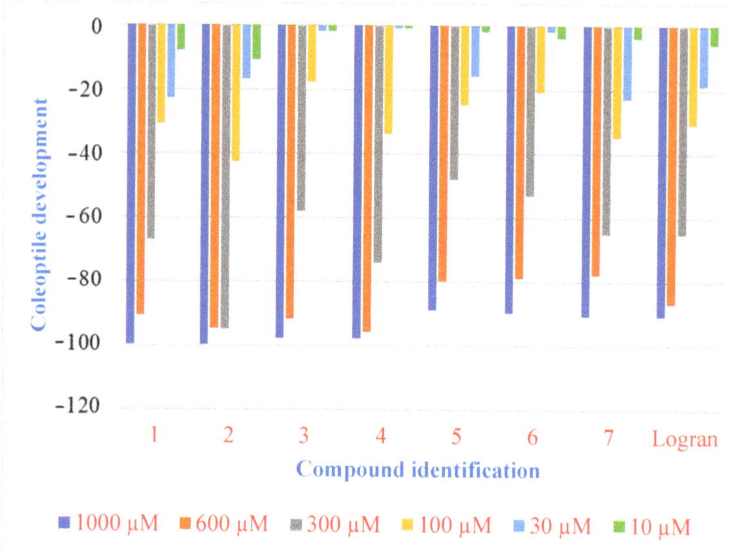

Figure 5. Bioactivity evaluation of compounds **1–7** and Logran on coleoptile development.

2.3. Herbicide Activity

In this study, the efficacy of compounds **2**, **4**, **5**, **6**, and **7** (Figure 6) in stimulating or inhibiting root and shoot development was evaluated using three plant species: *Allium cepa* (onion), *Lepidium sativum* (garden cress), and *Lactuca sativa* (lettuce). These species were selected due to their well-documented sensitivity to growth-regulating compounds, making them suitable bioindicators for assessing potential stimulatory or inhibitory effects.

The experimental process involved exposing seeds of each species to different concentrations of the selected compounds under controlled conditions. After the germination period, the seedlings were carefully analyzed and measured to determine variations in growth parameters. The collected data were then presented in graphical form, illustrating percentage differences compared to the control group.

In this analysis, the value of zero corresponds to the control, indicating no change in growth under water treatment. Positive percentage values denote stimulation, reflecting an increase in root or shoot development, while negative percentage values indicate inhibition, representing a reduction in growth compared to the control. These results provide insights into the bioactivity of the tested compounds, contributing to a better understanding of their potential applications in plant growth regulation [46,47].

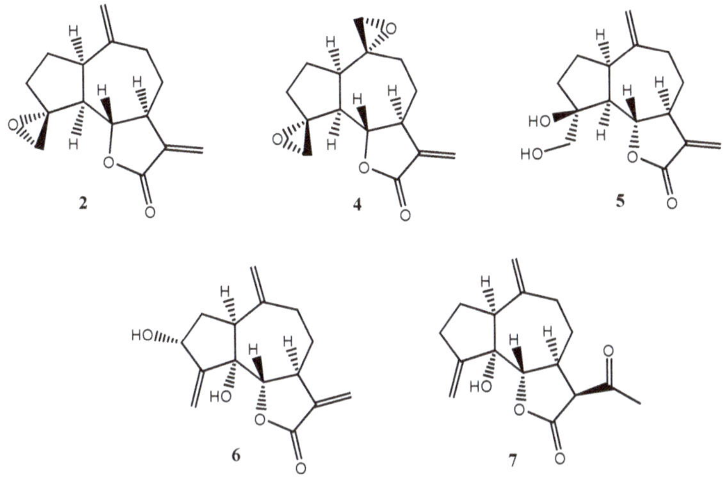

Figure 6. Compounds employed in the seed's bioassay.

As illustrated in Figure 7, compounds **4** and **6** demonstrated the most significant inhibitory effects on the shoot length of *A. cepa* when compared to the other tested compounds. The data indicate that compound **4** exhibits inhibitory effect, achieving 51% inhibition at a concentration of 100 µM and an even higher 60% inhibition at 60 µM, suggesting a potent dose-dependent response. Similarly, compound **6** also displayed notable activity, with an inhibition percentage of 40% at 100 µM. These findings highlight the promising potential of these compounds as effective plant growth inhibitors, reinforcing their role in herbicidal applications.

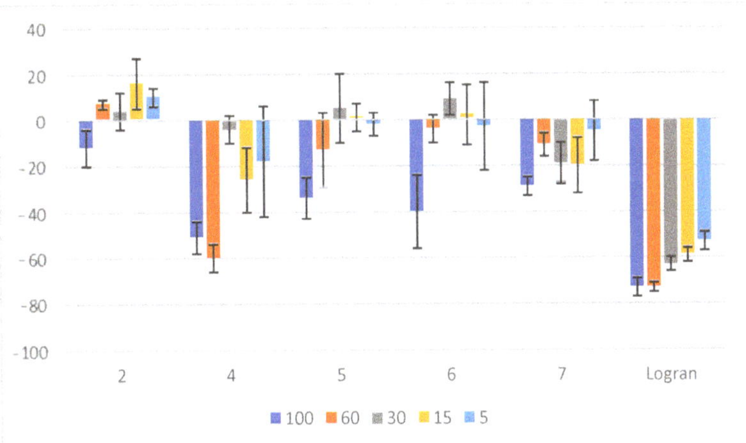

Figure 7. Effect of compounds **2** and **4–7** on the shoot length of *A. cepa*. Shoot length of seedlings from onion seeds exposed to aqueous 0.1% (*v/v*) DMSO solutions of compounds at different concentrations. Controls consisted of deionized water with the same concentration of DMSO. Values are expressed as percentage difference from the negative control, calculated as: shoot length (%) = [(length − length of negative control)/length of negative control] × 100. Error bars represent the standard deviation.

Regarding the inhibition of radicle length in *A. cepa*, compound **4** stands out as the most effective among all the tested compounds, exhibiting an impressive 80% inhibition at 100 µM and 77% inhibition at 60 µM (Figure 8). This indicates a strong inhibitory effect even at a slightly lower concentration. Furthermore, all other tested compounds also demonstrated notable inhibitory activity, with inhibition percentages exceeding 40% at 100 μM. These results emphasize the high efficacy of these compounds, particularly at a concentration of 100 μM, in suppressing root elongation.

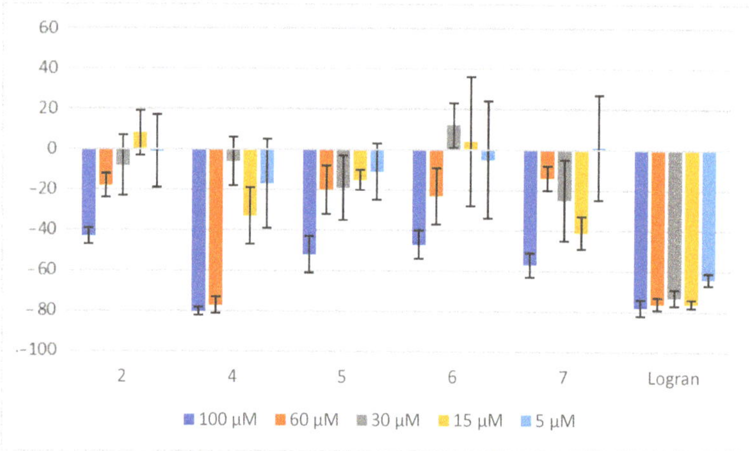

Figure 8. Effect of compounds **2** and **4–7** on the radicle length of *A. cepa*. Radicle length of seedlings from onion seeds exposed to aqueous 0.1% (*v/v*) DMSO solutions of compounds at different concentrations. Controls consisted of deionized water with the same concentration of DMSO. Values are expressed as percentage difference from the negative control, calculated as: shoot length (%) = [(length − length of negative control)/length of negative control] × 100. Error bars represent the standard deviation.

Compound **4** once again demonstrates remarkable biological activity, this time in the inhibition of *L. sativum* stalk length (Figure 9). At a concentration of 100 μM, this compound exhibits an impressive inhibition rate of 80%, which is twice as effective as the commercially available herbicide Logran, used as a reference compound in this study. This significant difference highlights the potential of compound **4** as a potent growth inhibitor.

As illustrated in Figure 10, all the tested compounds demonstrated a stimulatory effect on the growth of the radicle of garden cress. This suggests that, at certain concentrations, these compounds may play a role in promoting root elongation and overall seedling development. However, an exception was observed for compound **4**, which, at a concentration of 100 μM, significantly inhibited the root growth of *L. sativum* by 80%. This indicates a strong suppressive effect at this specific concentration, potentially interfering with root elongation and development. Such findings highlight the varying influence of different compounds on plant growth, depending on their concentration and the species under study.

No significant changes were observed for the compounds tested when compared to the length of the stem of *L. sativa*. However, upon further analysis, as illustrated in Figure 11, it was verified that compound **5** had a notable inhibitory effect on radicle growth. Specifically, at a concentration of 100 μM, compound **5** reduced radicle length by 28%, making it the most significant change observed among all tested compounds. This suggests that compound **5** exhibits a measurable impact on early seedling development, potentially affecting root elongation more than stem growth. Compounds **4**, **6**, and **7** presented stimulatory effect on the growth of lettuce. These three compounds stimulated almost 40% the growth of the radicle of *L. sativa*.

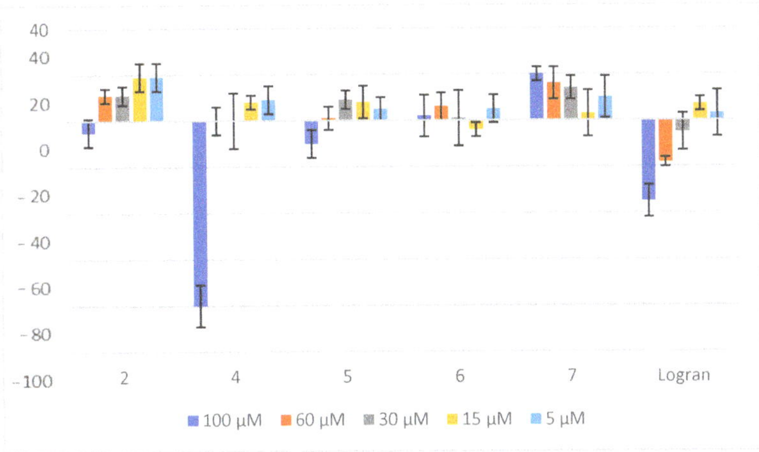

Figure 9. Effect of compounds **2** and **4–7** on the stalk length of *L. sativum*. Shoot length of seedlings of garden cress seeds exposed to aqueous 0.1% (*v/v*) DMSO solutions of the tested compounds at different concentrations. Controls consisted of deionized water with the same concentration of DMSO. Values are expressed as percentage difference from the negative control, calculated as: shoot length (%) = [(length − length of negative control)/length of negative control] × 100. Error bars represent the standard deviation.

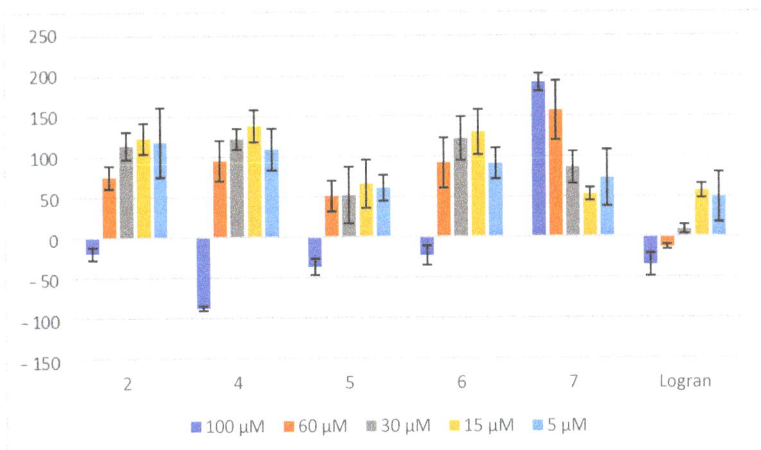

Figure 10. Effect of compounds **2** and **4–7** on the radicle length of *L. sativum*. Radicle length of seedlings from garden cress seeds exposed to aqueous 0.1% (*v/v*) DMSO solutions of compounds at different concentrations. Controls consisted of deionized water with the same concentration of DMSO. Values are expressed as percentage difference from the negative control, calculated as: shoot length (%) = [(length − length of negative control)/length of negative control] × 100. Error bars represent the standard deviation.

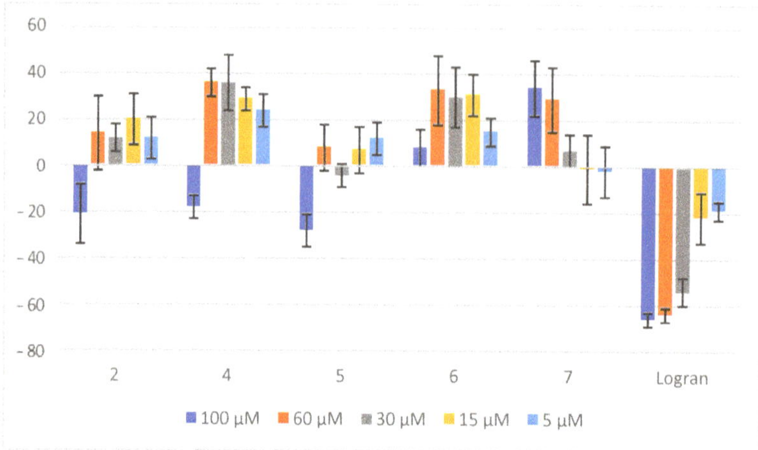

Figure 11. Effect of compounds **2** and **4–7** on the radicle length of *L. sativa*. Radicle length of seedlings from lettuce seeds exposed to aqueous 0.1% (*v/v*) DMSO solutions of compounds at different concentrations. Controls consisted of deionized water with the same concentration of DMSO. Values are expressed as percentage difference from the negative control, calculated as: shoot length (%) = [(length − length of negative control)/length of negative control] × 100. Error bars represent the standard deviation.

3. Materials and Methods

General experimental procedures. The *Saussurea lappa* extract (100 g) was subjected to chromatographic separation using silica gel column chromatography. The column used for this process measured 30 cm in height and 10 cm in diameter. The separation was carried out using a mobile phase consisting of hexane and ethyl acetate in a 95:5 (*v/v*) ratio, ensuring effective elution of the desired compounds. After fractionation and purification, pure dehydrocostus lactone (DHC) (5.0 g) was successfully isolated. The structural identity and purity of the isolated DHC were confirmed using various spectroscopic techniques, including ^{1}H and ^{13}C nuclear magnetic resonance (NMR) spectroscopy, as well as two-dimensional (2D) high-resolution NMR analyses (Table 1). The detailed spectral data supporting this identification are provided in Figures S2–S7 in the Supplementary Information (SI).

Table 1. Nuclear magnetic resonance data of DHC.

δ_H	Hydrogen	COSY	δ_C	Carbon
0.82–0.72	8′	8 × 8′, 9 × 8′,7 × 8′	30.7	8
1.60–1.46	2,8,9′	9′ × 14, 7 × 8, 8′ × 8, 9′ × 9, 1 × 2	30.4	2
1.98	9	8′, 9′, 7, 14	32.8	3
2.11	7	6 × 7,7 × 8, 7 × 8′, 7 × 13, 7 × 13′	36.4	9
2.41–2.20	1,3,3′,5 (m)	1 × 2, 1 × 2′, 3 × 2, 3 × 2′, 1 × 14	44.7	7
3.48	6 (t, *J* = 8 Hz)	6 × 7, 5 × 6	47.5	5
4.60	14,14′ (m)	1 × 14, 9 × 14, 9′ × 14	52.1	1
4.87	13′ (d, *J* = 4 Hz)	13′ × 7, 13 × 13′	84.4	6
5.07	15′ (m)	15′ × 5, 15′ × 3, 15′ × 3′	109.7	15
5.46	15 (m)	15 × 5, 15 × 3, 15 × 3′	112.0	14
6.11	13 (d, *J* = 4 Hz)	13 × 7, 13 × 13′	118.8	13
			140.7	4
			149.6	10
			151.4	11
			169.2	C=O

Epoxide derivatives—(3a*S*,9*S*,9b*S*)-3,6-dimethylenedecahydro-2*H*-spiro[azuleno[4,5-b]furan-9,2′-oxiran]-2-one **(1)**, (3a*S*,9*R*,9b*S*)-3-methylenedecahydro-2*H*-spiro[azuleno[4,5-b]furan-9,2′-oxiran]-2-one **(2)**, (3a*S*,6*R*,9b*S*)-3,9-dimethylenedecahydro-2*H*-spiro[azuleno[4,5-b]furan-6,2′-oxiran]-2-one **(3)**, (2*R*,3a′*S*,9′*R*,9b′*S*)-3′-methyleneoctahydrodispiro[oxirane-2,6′-azuleno[4,5-b]furan-9′,2″-oxiran]-2′(3′*H*)-one **(4)** and the diol (3a*S*,9*S*,9b*S*)-9-hydroxy-9-(hydroxymethyl)-6-methylenedecahydroazuleno[4,5-b]furan-2(3*H*)-one **(5)**—were prepared by epoxidation reaction of DHC with *meta*-chloroperbenzoic acid. All the structures were confirmed by high resolution NMR spectroscopy (the compounds were separated by HPLC, analytic column, refractive index detector, and characterized by ^{1}H and ^{13}C NMR, n.O.e experiments, HSQC and HMBC).

3.1. Synthetic Procedures

Epoxidation of DHC. *meta*-chloroperbenzoic acid (0.79 g, MW 172.57 g/mol, 4.58 mmol) in 15 mL anhydrous dichloromethane (DCM) was added to an ice cooled suspension of DHC (0.50 g, MW 230 g/mol, 2.17 mmol), anhydrous sodium acetate (0.46 g, MW 82.03 g/mol, 5.61 mmol), and powder molecular sieves (2 g) in anhydrous DCM (10 mL). The ice bath was removed, and the reaction mixture was stirred for 3 h. The mixture was fractioned through a neutral alumina column and eluted with DCM (100 mL) and a mixture of hexane–ethyl acetate 20% (500 mL). The fractions were combined and concentrated under reduced pressure to afford clear yellow oil (0.270 g). The oily residue (4 mg) was dissolved in hexane–ethyl acetate 60% (0.2 mL) and injected in the HPLC with semi-preparative column eluting with hexane–ethyl acetate 60%. The reaction yield was calculated employing the chromatogram obtained.

(3aS,9S,9bS)-3,6-dimethylenedecahydro-2H-spiro[azuleno[4,5-b] furan-9,2′-oxiran]-2-one (**1**)

$[\alpha]_D$ = 10.1 (methanol). IR $\bar{\upsilon}$ cm^{-1}: 2931, 1768, 1638, 1259, 1140, 999. ^{1}H NMR (400 MHz, C_6D_6): δ 6.06 (dd, *J* = 0.4, 2.8 Hz, H13), 4.81 (dd, *J* = 0.4, 2.8 Hz, H13′), 4.64 (d, *J* = 4.8 Hz, H14), 4.61 (d, *J* = 4.8 Hz, H14′), 3.31 (d, *J* = 4.8 Hz, H15), 3.23 (dd, *J* = 8.8, 11.0 Hz, H6), 2.76 (q, *J* = 8 Hz, H1), 2.57 (d, *J* = 4.8 Hz, H15′), 1.71–1.94 (m, H2, H3, H5, H7, H9), 1.38–1.56 (m, H2′, H3′, H8, H9′), 0.67–0.77 (m, H8′). ^{13}C NMR (100 MHz, C_6D_6): δ 28.7 (C2), 29.6 (C8), 31.7 (C3), 33.1 (C9), 45.5 (C7), 47.02 (C1), 49.6 (C15), 53.4 (C5), 65.8 (C4), 81.0 (C6), 113.5 (C14), 118.6 (C13), 140.5 (C10), 148.6 (C11), 168.7 (C=O).

(3aS,9R,9bS)-3-methylenedecahydro-2H-spiro[azuleno[4,5-b] furan-9,2′-oxiran]-2-one (**2**)

$[\alpha]_D$ = −20.0 (methanol). IR $\bar{\upsilon}$ cm^{-1}: 2930, 1767, 1635, 1260, 1130, 995. ^{1}H NMR (400 MHz, C_6D_6): δ 6.03 (d, *J* = 3.2 Hz, H13), 4.88 (m, H14), 4.82 (d, *J* = 3.2 Hz, H13′), 4.74 (m, H14′), 3.80–3.83 (m, H6, d, *J* =12 Hz, H15), 3.31 (d, *J* =12 Hz, H15′), 2.22–2.29 (m, H1), 2.12 (dt, *J*= 4, 12 Hz, H7), 1.87–1.96 (m, H5, H3′), 1.63–1.77 (m, H2′, H3), 1.40–1.54 (m, H2, H9, H9′), 1.29–1.36 (m, H8), 0.60–0.77 (m, H8′). ^{13}C NMR: (100 MHz, C_6D_6): δ 24.4 (C8), 26.7 (C2), 32.4 (C3), 35.5 (C9), 44.3 (C7), 45.4 (C1), 49.4 (C15), 50.3 (C5), 57.1 (C4), 83.4 (C6), 108.9 (C14), 118.9 (C13), 140.4 (C10), 151.4 (C11), 168.9 (C=O).

(3aS,6R,9bS)-3,9-dimethylenedecahydro-2H-spiro[azuleno[4,5-b]furan-6,2′-oxiran]-2-one (**3**)

$[\alpha]_D$ = −7.80 (methanol). IR $\bar{\upsilon}$ cm^{-1}: 2933, 1765, 1257, 1149, 999. ^{1}H NMR (400 MHz, C_6D_6): δ 5.32–5.33 (m, H15), 4.92–4.95 (m, H15′), 4.84 (d, *J* = 2.8 Hz, H13′), 3.41 (dd, *J* = 8, 10 Hz, H6), 2.32–2.37 (m, H5), 2.00–2.15 (m, H7, H3, H3′, H14), 1.88–1.96 (m, H1, H14′), 1.27–1.53 (m, H2, H2′, H8, H9), 1.12–1.18 (m, H9′), 0.59–0.69 (m, H8′). ^{13}C NMR (100 MHz, C_6D_6): δ 24.5 (C2), 26.6 (C8), 32.3 (C3), 35.4 (C9), 44.2 (C7), 45.3 (C1), 49.3 (C14), 50.5 (C5), 57.1 (C10), 83.4 (C6), 108.8 (C15), 119.1 (C13), 140.3 (C11), 151.4 (C4), 169.1 (C=O).

(2R,3a′S,9′R,9b′S)-3′-methyleneoctahydrodispiro[oxirane-2,6′-azuleno[4,5-b]furan-9′,2′-oxiran]-2′(3′H)-one (**4**)

$[\alpha]_D$ = 15.5 (methanol). IR $\bar{\upsilon}$ cm^{-1}: 2935, 1767, 1256, 1130, 999. ^{1}H NMR (400 MHz, C_6D_6): δ 6.10 (d, *J* = 2.8 Hz, H13), 4.82 (d, *J* = 2.8 Hz, H13′), 3.63 (dd, *J* = 9.2, 10 Hz, H6),

3.12 (d, J = 4.4 Hz, H15′), 2.71 (dd, J = 2, 4.4 Hz, H14), 2.35 (d, J = 4.8 Hz, H15′), 2.06–2.17 (m, H5, H14′, H15′), 1.86–1.91 (m, H1, H7), 1.54–1.65 (m, H2, H3), 1.36–1.49 (m, H2′, H9), 1.21–1.30 (m, H8), 1.12–1.18 (m, H3′, H9′), 0.50–0.61 (m, H8′). ^{13}C NMR (100 MHz, C_6D_6): δ 22.4 (C2), 27.1 (C8), 32.7 (C3), 38.6 (C9), 43.6 (C5), 44.1 (C7), 47.3 (C14), 47.4 (C1), 48.4 (C15), 56.2 (C10), 64.6 (C4), 80.4 (C6), 119.3 (C13), 139.5 (C11), 168.9 (C=O).

(3aS,9S,9bS)-9-hydroxy-9-(hydroxymethyl)-6-methylenedecahydroazuleno[4,5-b]furan-2(3H)-one (**5**)

$[\alpha]_D$ = 19.2 (methanol). IR $\overline{\upsilon}$ cm^{-1}: 3472, 2937, 1770, 1257, 1131, 996. ^{1}H NMR (400 MHz, C_6D_6): δ 4.07 (d, J = 3.6 Hz, H13), 4.89–4.90 (m, H14), 4.83 (d, J = 3.6 Hz, H13′), 4.77–4.78 (m, H14′), 3.88 (t, J = 9.6 Hz, H6), 3.01 (d, J = 4.8 Hz, H15), 2.36 (d, J = 4.8 Hz, H15′), 2.22–2.27 (m, H1), 1.96 (t, J = 9.2 Hz, H5), 1.74–1.85 (m, H2, H3, H9), 1.42–1.66 (m, H2′, H3′, H8, H9′), 0.71–0.81 (m, H8′). ^{13}C NMR (100 MHz, C_6D_6): δ 29.4 (C2), 31.0 (C8), 33.4 (C3), 38.5 (C9), 44.3 (C7), 46.3 (C1), 47.6 (C15), 49.0 (C5), 64.3 (C4), 81.0 (C6), 112.5 (C13), 118.9 (C14), 139.9 (C11), 149.3 (C10), 169.1 (C=O).

Synthesis of 5α-hydroxy-isozaluzanine C (**11**). Selenium oxide (2 eq. molar) and *tert*-butylhydroperoxide (2 eq. molar) were added to a solution of DHC (0.50 g, MW 230 g/mol, 2.17 mmol) in anhydrous DCM (50 ML) and stirred for 4 h under nitrogen atmosphere. Hydroxy-isozaluzanine C was fractioned in a column of silica gel and purified by HPLC (semi-preparative column; hexane–ethyl acetate 8:2) in 35% yield according to the previously reported conditions [48].

Synthesis of 13-oxo-epi-costuslactone. Aqueous sodium carbonate (20%) was added dropwise to a solution of DHC (0.200 g) in HMPA (20 mL) until it became turbid (around 16 mL). The reaction mixture was magnetically stirred and heated at 90 °C for 6 days. The mixture was extracted with ethyl acetate (4 × 10 mL) and the combined organic phases were washed with 2M aqueous hydrochloric acid (3 × 10 mL), 20% aqueous sodium carbonate (2 × 10 mL), and brine (2 × 10 mL). The organic phase was dried with anhydrous sodium sulphate, filtered, and concentrated under reduced pressure. The mixture of 13-hydroxycostuslactone and 13-hydroxy-epi-costuslactone was oxidized to the corresponding aldehyde using Dess–Martin periodinane according to the previously reported conditions [49].

3.2. Coleoptile Bioassay

Wheat seeds (*Triticum aestivum*) were sown in 15 cm diameter Petri dishes lined with moist filter paper and incubated in darkness at 25 °C for three days [50]. Coleoptiles measuring 25–35 mm in length were selected under a green safelight. A 3 mm section from the tip was removed and discarded, while the next 4 mm segment was used for the bioassay. After cutting, coleoptiles were soaked in distilled water for one hour before being randomly selected and placed in vials containing the test solutions. The commercial herbicide Logran (Novartis), containing 2-(2-chloroethoxy)-N-[[(4-methoxy-6-methyl-1,3,5-triazin-2-yl)amino]carbonyl] benzene-sulfonamide (Triasulfuron), was used as an internal positive reference.

Fractions were tested at concentrations of 1000, 600, 300, 100, 30, and 10 μM in a buffered nutritive aqueous solution (citric acid–sodium hydrogen phosphate buffer, pH 5.6; 2% sucrose). Stock solutions were prepared in DMSO and diluted with the buffer to achieve the desired concentrations, ensuring a final DMSO concentration of no more than 0.5% *v/v*. The control consisted of nutritive aqueous solution containing DMSO 0.5% *v/v*. Subsequent dilutions maintained the same buffer and DMSO concentrations.

Bioassays were conducted in 10 mL test tubes, with each tube containing five coleoptiles immersed in 2 mL of test solution. Three replicates were prepared for each test solution, and all experiments were performed in duplicate. The test tubes were placed in a roller tube apparatus and rotated at 6 rpm for 24 h at 22 °C in darkness. Results are displayed as percentage differences from the control in bar charts. A value of zero represents the control, while positive values indicate stimulation of the measured parameters, and negative values indicate inhibition [15].

3.3. *Seed Germination Bioassay*

The bioassays consisted of the germination of 25 seeds in the absence of light at 25 °C on 5 cm Petri plastic plates containing Whatman N° 1 filter paper, and 5 mL of the test or control solution. The stock solutions were prepared using DMSO (0.1% *v/v*) and the test solutions were obtained from the dilution of that stock solution. The commercial herbicide Logran (Novartis), containing 2-(2-chloroethoxy)-*N*-[[(4-methoxy-6-methyl-1,3,5-triazin-2-yl)amino]carbonyl] benzenesulfonamide (Triasulfuron), was used as an internal reference.

Controls consisted of deionized water containing 0.1% (*v/v*) DMSO. Three repetitions of each treatment were carried out with concentrations of 100, 60, 30, 15, 5 m*M* and control. The species tested were as follows (incubation time between parentheses): *Allium cepa* L. (onion, monocotiledone, 5 days), *Lactuca sativa* L. (lettuce, dicotyledone, 5 days), and *Lepidum sativum* L. (garden cress, dicotyledone, 3 days). The statistical analysis of the data was carried out using Welch's test. The significance levels were set at 0.01 and 0.05, meaning that differences were considered statistically significant if the probability of obtaining the observed results by chance was less than 1% ($p < 0.01$) or 5% ($p < 0.05$) [51].

4. Conclusions

In this study, five novel compounds were successfully synthesized from dehydrocostus lactone, a bioactive sesquiterpene lactone known for its diverse biological properties. A total of seven compounds were systematically evaluated for their coleoptile activity, assessing their potential effects on the elongation and growth of the coleoptile, an essential structure in seedling development. Additionally, five of these compounds were further tested for their herbicidal activity against three distinct plant species: *Allium cepa (onion)*, *Lepidium sativum (garden cress)*, and *Lactuca sativa (lettuce)*. These plant species were selected due to their rapid germination and sensitivity to growth inhibitors, making them ideal models for herbicidal screening.

Among the lactone derivatives evaluated in this study, compound **4**, which was synthesized through the epoxidation of dehydrocostus lactone (DHC), emerged as the most potent inhibitor of plant growth. This compound demonstrated the highest inhibitory activity, significantly reducing the growth of *L. sativum*. Specifically, it exhibited an 80% inhibition of stalk length and an even greater 88% inhibition of radicle length of garden cress, indicating a strong growth-suppressing effect. In contrast, the majority of the other tested compounds stimulated the growth of *L. sativum*, further emphasizing the unique and powerful inhibitory nature of compound **4**. These findings suggest that this compound holds promising potential as a natural herbicidal agent, capable of effectively limiting plant development.

A possible explanation for the enhanced biological activity of compound **4** lies in the presence of two epoxide functional groups within its molecular structure. Epoxides are well known for their high electrophilicity, making them highly reactive toward nucleophilic attack by biomolecules, including plant enzymes. This reactivity allows epoxides to interact with key enzymatic pathways involved in plant growth and development, potentially leading to enzyme inhibition or disruption of metabolic processes essential for cell division and elongation. Consequently, the presence of two epoxide groups in compound **4** may significantly enhance its herbicidal potency, as it provides multiple reactive sites that increase its likelihood of interfering with plant biochemical mechanisms. This structural feature likely accounts for the superior inhibitory effects observed in the growth of *L. sativum* when compared to other tested compounds.

Future studies should focus on evaluating the stability of the synthesized compounds in environmental conditions, particularly in soil and water, to assess their biodegradability. Understanding the degradation pathways and persistence of these compounds is essential for determining their environmental impact and ensuring their suitability for agricultural applications. Additionally, bioassays should be conducted to assess the effects of the synthesized compounds on beneficial organisms, such as natural enemies of pests and pollinators, including ladybirds and bees. Evaluating their impact on these non-target species is crucial for determining the ecological safety of the compounds.

Supplementary Materials: The following supporting information can be downloaded at: https://www.mdpi.com/article/10.3390/plants14071111/s1, Figures S1–S40; Tables S1–S11.

Author Contributions: Conceptualization, J.C.G.G. and J.M.G.M.; methodology, J.C.G.G.; validation, E.S.A.; formal analysis, E.S.A.; investigation, E.S.A.; resources, F.A.M.; data curation, F.A.M.; writing—original draft preparation, S.S.F.; writing—review and editing, E.S.A.; visualization, J.M.G.M.; supervision, J.C.G.G.; funding acquisition, E.S.A. and F.A.M. All authors have read and agreed to the published version of the manuscript.

Funding: This research was supported by Fundação de Amparo à Pesquisa do Estado de Minas Gerais—FAPEMIG, Conselho Nacional de Desenvolvimento Científico e Tecnológico—CNPq and Coordenação de Aperfeiçoamento de Pessoal de Nível Superior—CAPES.

Data Availability Statement: Data is contained within the article and Supplementary Materials.

Acknowledgments: We would like to thank Universidade Federal de Viçosa, Universidad de Cádiz, and the Department of Education of the Andalusian Government.

Conflicts of Interest: The authors declare no conflicts of interest.

References

1. Cárdenas, D.M.; Bajsa-Hirschel, J.; Cantrell, C.L.; Rial, C.; Varela, R.M.; Molinillo, J.M.G.; Macías, F.A. Evaluation of the Phytotoxic and Antifungal Activity of C17-Sesquiterpenoids as Potential Biopesticides. *Pest Manag. Sci.* **2022**, *78*, 4240–4251. [CrossRef] [PubMed]
2. Miranda, D.A.; Santos, R.T.D.S.; Bacha, A.L.; Rodrigues, J.D.S.; Alves, P.L.d.C.A.; Kuva, M.A. Estudo de Seleção Da Comunidade Infestante Por Herbicidas Utilizando Técnicas de Análise Multivariada. *Rev. Bras. De Herbic.* **2020**, *19*, 1–13. [CrossRef]
3. He, B.; Hu, Y.; Wang, W.; Yan, W.; Ye, Y. The Progress towards Novel Herbicide Modes of Action and Targeted Herbicide Development. *Agronomy* **2022**, *12*, 2792. [CrossRef]
4. Qu, R.; He, B.; Yang, J.; Lin, H.; Yang, W.; Wu, Q.; Li, Q.X.; Yang, G. Where Are the New Herbicides? *Pest Manag. Sci.* **2021**, *77*, 2620–2625. [CrossRef]
5. Fraga, B.M.; Díaz, C.E.; Bailén, M.; González-Coloma, A. Sesquiterpene Lactones from Artemisia Absinthium. Biotransformation and Rearrangement of the Insect Antifeedant 3α-Hydroxypelenolide. *Plants* **2021**, *10*, 891. [CrossRef]
6. Sparks, T.C.; Bryant, R.J. Impact of Natural Products on Discovery of, and Innovation in, Crop Protection Compounds. *Pest Manag. Sci.* **2022**, *78*, 399–408. [CrossRef]
7. Cárdenas, D.; Rial, C.; Varela, R.; Molinillo, J.; Macías, F. Synthesis of Pertyolides A, B, and C: A Synthetic Procedure to C 17 -Sesquiterpenoids and a Study of Their Phytotoxic Activity. *J. Nat. Prod.* **2021**, *84*, 2295–2302. [CrossRef]
8. Zhou, C.; Luo, X.; Chen, N.; Zhang, L.; Gao, J. C–P Natural Products as Next-Generation Herbicides: Chemistry and Biology of Glufosinate. *J. Agric. Food Chem.* **2020**, *68*, 3344–3353. [CrossRef]
9. Paço, A.; Brás, T.; Santos, J.O.; Sampaio, P.; Gomes, A.C.; Duarte, M.F. Anti-Inflammatory and Immunoregulatory Action of Sesquiterpene Lactones. *Molecules* **2022**, *27*, 1142. [CrossRef]
10. Brás, T.; Neves, L.A.; Crespo, J.G.; Duarte, M.F. Advances in Sesquiterpene Lactones Extraction. *TrAC Trends Anal. Chem.* **2023**, *158*, 116838. [CrossRef]

11. Seregheti, T.M.Q.; Pinto, A.P.R.; Gonçalves, M.d.C.; Antunes, A.S.; Almeida, W.A.d.S.; Machado, R.S.; Silva, J.N.; Ferreira, P.M.P.; Pessoa, C.; Dos Santos, V.M.R.; et al. Antiproliferative and Photoprotective Activities of the Extracts and Compounds from Calea Fruticosa. *Braz. J. Med. Biol. Res.* **2020**, *53*, 1–8. [CrossRef]
12. Neverova, N.A.; Zhabayeva, A.N.; Levchuk, A.A.; Babkin, V.A.; Beisenbaev, A.R.; Larina, L.I.; Sapozhnikov, A.N.; Adekenov, C.M. A Study of the Physicochemical Properties of Mechanically Treated Arglabin and Its Mechanocomposites Based on Arabinogalactan. *Russ. J. Bioorg. Chem.* **2020**, *46*, 1358–1363. [CrossRef]
13. Colasurdo, D.D.; Arancibia, L.A.; Naspi, M.L.; Laurella, S.L. Using DP4+ Probability for Structure Elucidation of Sesquiterpenic Lactones: The Case of (−)-Istanbulin A. *J. Phys. Org. Chem.* **2022**, *35*, e4282. [CrossRef]
14. Wang, J.; Zheng, Q.; Wang, H.; Shi, L.; Wang, G.; Zhao, Y.; Fan, C.; Si, J. Sesquiterpenes and Sesquiterpene Derivatives from Ferula: Their Chemical Structures, Biosynthetic Pathways, and Biological Properties. *Antioxidants* **2024**, *13*, 7.
15. Moraes, F.C.; Alvarenga, E.S.; Amorim, K.B.; Demuner, A.J.; Pereira-Flores, M.E. Novel Platensimycin Derivatives with Herbicidal Activity. *Pest Manag. Sci.* **2016**, *72*, 580–584. [CrossRef]
16. Kadhila, N.P.; Sekhoacha, M.; Tselanyane, M.; Chinsembu, K.C.; Molefe-Khamanga, D.M. Determination of the Antiplasmodial Activity, Cytotoxicity and Active Compound of Pechuel-Loeschea Leubnitziae O. Hoffm. (Asteraceae) of Namibia. *SN Appl. Sci.* **2020**, *2*, 1328. [CrossRef]
17. Agatha, O.; Mutwil-Anderwald, D.; Tan, J.Y.; Mutwil, M. Plant Sesquiterpene Lactones. *Phil. Trans. R. Soc. B* **2024**, *379*, 20230350. [CrossRef]
18. Kovács, B.; Hohmann, J.; Csupor-Löffler, B.; Kiss, T.; Csupor, D. A Comprehensive Phytochemical and Pharmacological Review on Sesquiterpenes from the Genus Ambrosia. *Heliyon* **2022**, *8*, e09884.
19. Wu, W.; Huang, H.; Su, J.; Yun, X.; Zhang, Y.; Wei, S.; Huang, Z.; Zhang, C.; Bai, Q. Dynamics of Germination Stimulants Dehydrocostus Lactone and Costunolide in the Root Exudates and Extracts of Sunflower. *Plant Signal. Behav.* **2022**, *17*, 2025669. [CrossRef]
20. Laber, L.; Eichberg, C.; Zimmerbeutel, A.; Düring, R.A.; Donath, T.W. Effects of Macrocyclic Lactone Anthelmintics on Seed Germination of Temperate Grassland Species. *Plant Biol.* **2023**, *25*, 1046–1057. [CrossRef]
21. Zorrilla, J.G.; Cárdenas, D.M.; Rial, C.; Molinillo, J.M.G.; Varela, R.M.; Masi, M.; Macías, F.A. Bioprospection of Phytotoxic Plant-Derived Eudesmanolides and Guaianolides for the Control of *Amaranthus Viridis, Echinochloa Crus-Galli*, and *Lolium Perenne* Weeds. *J. Agric. Food Chem.* **2024**, *72*, 1797–1810. [CrossRef] [PubMed]
22. Kalisz, S.; Kivlin, S.N.; Bialic-Murphy, L. Allelopathy Is Pervasive in Invasive Plants. *Biol. Invasions* **2021**, *23*, 367–371. [CrossRef]
23. McCoy, R.M.; Widhalm, J.R.; McNickle, G.G. Allelopathy as an Evolutionary Game. *Plant Direct* **2022**, *6*, e382. [CrossRef]
24. Schepetkin, I.A.; Kirpotina, L.N.; Mitchell, P.T.; Kishkentaeva, A.S.; Shaimerdenova, Z.R.; Atazhanova, G.A.; Adekenov, S.M.; Quinn, M.T. The Natural Sesquiterpene Lactones Arglabin, Grosheimin, Agracin, Parthenolide, and Estafiatin Inhibit T Cell Receptor (TCR) Activation. *Phytochemistry* **2018**, *146*, 36–46. [CrossRef]
25. Csuk, R.; Heinold, A.; Siewert, B.; Schwarz, S.; Barthel, A.; Kluge, R.; Ströhl, D. Synthesis and Biological Evaluation of Antitumor-Active Arglabin Derivatives. *Arch. Pharm.* **2012**, *345*, 215–222. [CrossRef]
26. Salin, A.V.; Shabanov, A.A.; Khayarov, K.R.; Islamov, D.R.; Voloshina, A.D.; Amerhanova, S.K.; Lyubina, A.P. Phosphine-Catalyzed Synthesis and Cytotoxic Evaluation of Michael Adducts of the Sesquiterpene Lactone Arglabin. *ChemMedChem* **2024**, *19*, e202400045. [CrossRef]
27. Huo, J.; Yang, S.P.; Ding, J.; Yue, J.M. Cytotoxic Sesquiterpene Lactones from Eupatorium Lindleyanum. *J. Nat. Prod.* **2004**, *67*, 1470–1475. [CrossRef]
28. Silverstein, R.M.; Webster, F.X.; Bryce, D.J. *Spectrometric Identification of Organic Compounds*, 8th ed.; John Wiley & Sons, Inc.: Hoboken, NJ, USA, 2015; ISBN 9780470616376.
29. Hernández-Guadarrama, A.; Díaz-Román, M.A.; Linzaga-Elizalde, I.; Domínguez-Mendoza, B.E.; Aguilar-Guadarrama, A.B. In Silico Analysis: Anti-Inflammatory and α-Glucosidase Inhibitory Activity of New α-Methylene-γ-Lactams. *Molecules* **2024**, *29*, 1973. [CrossRef]
30. Van Vu, Q.; Sayama, S.; Ando, M.; Kataoka, T. Sesquiterpene Lactones Containing an α-Methylene-γ-Lactone Moiety Selectively Down-Regulate the Expression of Tumor Necrosis Factor Receptor 1 by Promoting Its Ectodomain Shedding in Human Lung Adenocarcinoma A549 Cells. *Molecules* **2024**, *29*, 1866. [CrossRef]
31. Cala, A.; Zorrilla, J.G.; Rial, C.; Molinillo, J.M.G.; Varela, R.M.; Macías, F.A. Easy Access to Alkoxy, Amino, Carbamoyl, Hydroxy, and Thiol Derivatives of Sesquiterpene Lactones and Evaluation of Their Bioactivity on Parasitic Weeds. *J. Agric. Food Chem.* **2019**, *67*, 10764–10773. [CrossRef]
32. Bruno, M.; Rosselli, S.; Maggio, A.; Raccuglia, R.A.; Bastow, K.F.; Lee, K.H. Cytotoxic Activity of Some Natural and Synthetic Guaianolides. *J. Nat. Prod.* **2005**, *68*, 1042–1046. [CrossRef]
33. Wu, F.; Gao, Y.; Yang, W.; Sui, N.; Zhu, J. Biological Functions of Strigolactones and Their Crosstalk With Other Phytohormones. *Front. Plant Sci.* **2022**, *13*, 821563.

34. Mansoor, S.; Mir, M.A.; Karunathilake, E.M.B.M.; Rasool, A.; Ştefănescu, D.M.; Chung, Y.S.; Sun, H.J. Strigolactones as Promising Biomolecule for Oxidative Stress Management: A Comprehensive Review. *Plant Physiol. Biochem.* **2024**, *206*, 108282.
35. Kapoor, R.T.; Alam, P.; Chen, Y.; Ahmad, P. Strigolactones in Plants: From Development to Abiotic Stress Management. *J. Plant Growth Regul.* **2024**, *43*, 903–919. [CrossRef]
36. Jamil, M.; Alagoz, Y.; Wang, J.Y.; Chen, G.T.E.; Berqdar, L.; Kharbatia, N.M.; Moreno, J.C.; Kuijer, H.N.J.; Al-Babili, S. Abscisic Acid Inhibits Germination of Striga Seeds and Is Released by Them Likely as a Rhizospheric Signal Supporting Host Infestation. *Plant J.* **2024**, *117*, 1305–1316. [CrossRef]
37. Cao, K.; Qian, W.; Xu, Y.; Zhou, Z. Purification of Sesquiterpenes from Saussurea Lappa Roots by High Speed Counter Current Chromatography. *Iran. J. Pharm. Res.* **2019**, *18*, 1499–1507. [CrossRef]
38. Yang, M.; Zhang, J.; Li, Y.; Han, X.; Gao, K.; Fang, J. Bioassay-Guided Isolation of Dehydrocostus Lactone from Saussurea Lappa: A New Targeted Cytosolic Thioredoxin Reductase Anticancer Agent. *Arch. Biochem. Biophys.* **2016**, *607*, 20–26. [CrossRef]
39. Mori, K.; Matsui, M.; Sumiki, Y. Biochemical Studies on "Bakanae" Fungus. Part 55 Synthesis of Gibberellins: Part II. Infrared Spectra of the Lactones of Cyclohexane Series. *Agric. Biol. Chem.* **1961**, *25*, 205–222. [CrossRef]
40. Blank, D.E.; Demuner, A.J.; Baía, V.C.; Alvarenga, E.S.; Cerceau, C.I.; de Moura, N.F. Structural Elucidation of a Lactam Isolated from Bunchosia Glandulifera and Antioxidant Activity. *Peer Rev.* **2023**, *5*, 1–14. [CrossRef]
41. Kolmer, A.; Edwards, L.J.; Kuprov, I.; Thiele, C.M. Conformational Analysis of Small Organic Molecules Using NOE and RDC Data: A Discussion of Strychnine and α-Methylene-γ-Butyrolactone. *J. Magn. Reson.* **2015**, *261*, 101–109. [CrossRef]
42. Macías, F.A.; Alvarenga, E.S.; Galindo, J.C.G.; Molinillo, J.M.G.; Cerceau, C.I. Microwave-Assisted Rearrangement of Costunolide Catalyzed by Palladium(II). *J. Braz. Chem. Soc.* **2024**, *35*. [CrossRef]
43. Cala, A.; Salcedo, J.R.; Torres, A.; Varela, R.M.; Molinillo, J.M.G.; Macías, F.A. A Study on the Phytotoxic Potential of the Seasoning Herb Marjoram (*Origanum majorana* L.) Leaves. *Molecules* **2021**, *26*, 3356. [CrossRef] [PubMed]
44. Durán, A.G.; Calle, J.M.; Butrón, D.; Pérez, A.J.; Macías, F.A.; Simonet, A.M. Steroidal Saponins with Plant Growth Stimulation Effects; Yucca Schidigera as a Commercial Source. *Plants* **2022**, *11*, 3378. [CrossRef] [PubMed]
45. Rodríguez-Mejías, F.J.; Mottaghipisheh, J.; Schwaiger, S.; Kiss, T.; Csupor, D.; Varela, R.M.; Macías, F.A. Allelopathic Studies with Furanocoumarins Isolated from Ducrosia Anethifolia. In Vitro and in Silico Investigations to Protect Legumes, Rice and Grain Crops. *Phytochemistry* **2023**, *215*, 113838. [CrossRef]
46. Sangi, D.P.; Meira, Y.G.; Moreira, N.M.; Lopes, T.A.; Leite, M.P.; Pereira-Flores, M.E.; Alvarenga, E.S. Benzoxazoles as Novel Herbicidal Agents. *Pest Manag. Sci.* **2019**, *75*, 262–269. [CrossRef]
47. Teixeira, M.G.; Alvarenga, E.S.; Lopes, D.T.; Oliveira, D.F. Herbicidal Activity of Isobenzofuranones and in Silico Identification of Their Enzyme Target. *Pest Manag. Sci.* **2019**, *75*, 3331–3339. [CrossRef]
48. Macías, F.A.; Galindo, J.C.G.; Massanet, G.M. Potential Allelopathic Activity of Several Sesquiterpene Lactone Models. *Phytochemistry* **1992**, *31*, 1969–1977. [CrossRef]
49. Macías, F.A.; García-Díaz, M.D.; Pérez-de-Luque, A.; Rubiales, D.; Galindo, J.C.G. New Chemical Clues for Broomrape-Sunflower Host–Parasite Interactions: Synthesis of Guaianestrigolactones. *J. Agric. Food Chem.* **2009**, *57*, 5853–5864. [CrossRef]
50. Elakbawy, W.M.; Shanab, S.M.M.; Shalaby, E.A. Enhancement of Plant Growth Regulators Production from Microalgae Cultivated in Treated Sewage Wastewater (TSW). *BMC Plant Biol.* **2022**, *22*, 377. [CrossRef]
51. Macías, F.A.; Simonet, A.M.; Galindo, J.C.G. Bioactive Steroids and Triterpenes from Melilotus Messanensis and Their Allelopathic Potential. *J. Chem. Ecol.* **1997**, *23*, 1781–1803. [CrossRef]

plants

MDPI

Article

Anticancer Potential of *Cymbopogon citratus* L. Essential Oil: In Vitro and In Silico Insights into Mitochondrial Dysfunction and Cytotoxicity in Cancer Cells

Tamara Maksimović [1,2,†], Daliana Minda [3,†], Codruța Șoica [1,2], Alexandra Mioc [1,2,*], Marius Mioc [2,4], Daiana Colibășanu [2,4], Alexandra Teodora Lukinich-Gruia [5], Maria-Alexandra Pricop [5,6], Calin Jianu [7] and Armand Gogulescu [8]

1 Department of Pharmacology-Pharmacotherapy, Faculty of Pharmacy, "Victor Babes" University of Medicine and Pharmacy, Eftimie Murgu Square, No. 2, 300041 Timișoara, Romania; tamara.maksimovic@umft.ro (T.M.); codrutasoica@umft.ro (C.Ș.)
2 Research Center for Experimental Pharmacology and Drug Design (X-Pharm Design), "Victor Babes" University of Medicine and Pharmacy, Eftimie Murgu Square, No. 2, 300041 Timișoara, Romania; marius.mioc@umft.ro (M.M.); daiana.handa@umft.ro (D.C.)
3 Department Pharmacognosy-Phytotherapy, Faculty of Pharmacy, "Victor Babes" University of Medicine and Pharmacy, 2 Eftimie Murgu, 300041 Timisoara, Romania; daliana.minda@umft.ro
4 Department of Pharmaceutical Chemistry, Faculty of Pharmacy, "Victor Babes" University of Medicine and Pharmacy, Eftimie Murgu Square, No. 2, 300041 Timișoara, Romania
5 OncoGen Centre, Clinical County Hospital "Pius Branzeu", Blvd. Liviu Rebreanu 156, 300723 Timisoara, Romania; alexandra.gruia@hosptm.ro (A.T.L.-G.); alexandra.pricop@oncogen.ro (M.-A.P.)
6 Department of Applied Chemistry and Environmental Engineering and Inorganic Compounds, Faculty of industrial Chemistry, Biotechnology and Environmental Engineering, Polytechnic University of Timisoara, Vasile Pârvan 6, 300223 Timisoara, Romania
7 Faculty of Food Engineering, Banat's University of Agricultural Sciences and Veterinary Medicine "King Michael I of Romania" from Timisoara, Calea Aradului 119, 300629 Timișoara, Romania; calin.jianu@gmail.com
8 Department XVI: Balneology, Medical Rehabilitation and Rheumatology, "Victor Babes" University of Medicine and Pharmacy, 2 Eftimie Murgu, 300041 Timisoara, Romania; gogulescu.armand@umft.ro
* Correspondence: alexandra.mioc@umft.ro
† Authors with equal contribution jointly sharing the first author position.

Abstract: This study aims to assess the potential anticancer activity of lemongrass essential oil (LEO) using in vitro and in silico methods. The steam hydrodistillation of the aerial parts yielded 3.2% (wt) LEO. The GC-MS analysis of the LEO revealed the presence of α-citral (37.44%), β-citral (36.06%), linalool acetate (9.82%), and d-limonene (7.05%) as major components, accompanied by several other minor compounds. The antioxidant activity, assessed using the DPPH assay, revealed that LEO exhibits an IC_{50} value of 92.30 μg/mL. The cytotoxic effect of LEO, as well as LEO solubilized with Tween-20 (LEO-Tw) and PEG-400 (LEO-PEG), against a series of cancer cell lines (A375, RPMI-7951, MCF-7, and HT-29) was assessed using the Alamar Blue assay; the results revealed a high cytotoxic effect against all cell lines used in this study. Moreover, neither one of the tested concentrations of LEO, LEO-PG, or LEO-TW significantly affected the viability of healthy HaCaT cells, thus showing promising selectivity characteristics. Furthermore, LEO, LEO-PG, and LEO-TW increased ROS production and decreased the mitochondrial membrane potential (MMP) in all cancer cell lines. Moreover, LEO treatment decreased all mitochondrial respiratory rates, thus suggesting its ability to induce impairment of mitochondrial function. Molecular docking studies revealed that LEO anticancer activity, among other mechanisms, could be attributed to PDK1 and PI3Kα, where the major contributors are among the minor components of the essential oil. The highest active theoretical inhibitor against both proteins was β-caryophyllene oxide.

Keywords: lemongrass (*Cymbopogon citratus* L.) essential oil; anticancer activity; natural compounds; molecular docking

1. Introduction

Anticancer treatments have advanced significantly in recent decades, with novel emerging therapies triggered by the inherent challenges to conventional agents such as multidrug cell resistance and severe side effects [1]. Although providing superior clinical outcomes, such new strategies come with high costs and challenges of their own (i.e., ability to evade the immune response, development of efficient and biocompatible delivery systems, etc.) [2]. Given that cancer retains high incidence and mortality worldwide, making it a leading cause of death [3], there is continuous pressure to develop new agents able to provide more effective and selective alternatives for cancer treatment. Plants have provided numerous molecular scaffolds with anticancer efficacy but also various compounds valuable as adjuvants during chemotherapy, able to synergize with antitumor drugs in terms of acting as drug sensitizers, reducing resistance to drugs or alleviating drug side effects [4].

Essential oils are mixtures of volatile secondary metabolites of plant origin that have developed as part of the plant's chemical defense system against external threats such as herbivores, insects, or microorganisms, and they exhibit intrinsic pharmacological properties. They have been used extensively for the prevention and treatment of various diseases, including cancer [5]. Essential oils display a complex chemical composition containing up to 300 different molecules, among which phenols, alcohols, and aldehydes were found to induce the strongest antitumor properties by means of various mechanisms including cancer prevention, direct effects against cancer cells, and the tumor microenvironment [6].

Lemongrass essential oil (LEO) from *Cymbopogon flexuosus* was tested as an anticancer agent against colon, lung, cervix, oral, prostate, and leukemia cancer cells and showed strong cytotoxic effects through dose-dependent apoptosis; the in vitro effects correlated with the in vivo dose-dependent inhibition of tumor growth in the animal models of solid tumors [7]. LEO, as well as its main constituents citral and geraniol, were identified as effective antiproliferative agents in MCF7 breast cancer cells through the inhibition of HSP90 chaperone protein [8]; additionally, citral exhibits selective cytotoxic activity against breast cancer cells [9]. Citral induces multiple anticancer molecular mechanisms that include the accumulation of ROS in cancer cells with subsequent DNA damages and the inhibition of tubulin polymerization and aldehyde dehydrogenase isoform ALDH1A3, which favors cancer proliferation and chemoresistance [10]; moreover, other minor components in lemongrass essential oil induce cytotoxic effects through various mechanisms. However, despite the numerous papers reporting the direct correlation between the anticancer activity of LEO and its citral content, one study published in 2020 by Viktorová et al. revealed that neither the anticancer nor the antimicrobial activity of LEO are caused by the presence of citral; the authors concluded that the natural mixture of compounds identified in LEO presumably induce synergistic effects, and their use as a whole is more beneficial that the use of pure citral [11].

The current study aims to assess the potential of LEO, extracted from *Cymbopogon citratus* L., to act as an anticancer agent against a series of cancer cell lines (A375, RPMI-7951, MCF-7, and HT-29); the comparative assessment with pure citral was conducted as well. Additionally, the underlying molecular mechanisms were investigated by in silico methods.

2. Results

2.1. Chemical Composition of LEO

The steam hydrodistillation of the aerial parts of *Cymbopogon citratus* L. yielded 3.2% (wt) of a pale yellow to light greenish-yellow oil with a characteristic citrus-like lemony odor. The GC-MS analysis of the LEO revealed the presence of α-citral (37.44%), β-citral (36.06%), linalyl acetate (9.82%), d-limonene (7.05%), and β-caryophyllene (6.73%) as major components, accompanied by several other minor compounds, as shown in Table 1.

Table 1. Chemical composition of LEO determined through GC-MS.

No	Compound Name	RT	RI Calc	AI	Area % Calc
1	Terpinolene	5.61	975		0.05
2	d-Limonene	5.96	994	1024	7.05
3	Eucalyptol	6.19	1005		0.12
4	*p*-Cymene	7.28	1063	1020	0.23
5	6-Methyl-5-hepten-2-one	8.61	1133		1.20
6	Citronellal	11.45	1283	1148	2.06
7	α-Cubebene	11.76	1299	1387	0.18
8	1,2,5,5-Tetramethyl-1,3-cyclopentadiene	12.07	1316		0.11
9	Triluoroacetyl-lavandulol	12.22	1324		0.18
10	Ethenyl-cyclohexane	12.74	1351		0.27
11	Linalyl acetate	13.30	1381	1254	9.82
12	β-Caryophyllene	13.86	1410	1418	6.73
13	Humulene	15.27	1484		0.93
14	β-Citral	15.53	1498	1316	36.06
15	α-Citral	16.48	1548	1338	37.44
16	Neryl acetate	16.80	1565	1359	2.43
17	β-Caryophyllene oxide	20.83	1778		0.66

2.2. Determination of the Antioxidant Activity of LEO

LEO antioxidant scavenging activity was determined by employing 2,2-diphenyl-1-picrylhydrazyl (DPPH) and [2,2′-azino-bis(3-ethylbenzthiazoline-6-sulfonic acid)] (ABTS) radical scavenging assays. Pure ascorbic acid (AA) and butylated hydroxyanisole (BHA) were used as positive controls in the DPPH assay. For the ABTS assay, 1 mg/mL solutions of AA and BHA were used as standard references to determine percentage inhibition. The results are presented in Table 2.

Table 2. Scavenging antioxidant activity of LEO determined by DPPH and ABTS scavenging assays.

Sample	IC_{50} (μg/mL)	ABTS (Inh%)
LEO	92.30 ± 18.00	78.20 ± 0.040
AA	12.67 ± 5.28	88.84 ± 0.002
BHA	6.36 ± 13.56	88.79 ± 0.002

2.3. LEO Effect on Cell Viability

LEO presented diverse effects depending on the cell line and tested concentration. The HaCaT cells did not show a significant decrease in viability after LEO, LEO-PEG, and LEO-Tw treatments, even at the highest tested concentrations (200 μg/mL) (Figure 1A). The viability of all the cell lines treated for 48 h with PEG-400 and Tween 20 alone did not significantly differ from the control. Meanwhile, cellular viability was significantly affected in the cancer cell lines. A dose-dependent drop in viability was observed in the case of A375 cells, in which stimulation with 100 μg/mL and 200 μg/mL LEO resulted in 32.3 ± 6.1% and 16.5 ± 5.9% cell viability vs. control (100%); treatment with 100 μg/mL and 200 μg/mL LEO-PEG decreased the viability to 19.7 ± 12.3% and 14.6 ± 3.1%; and treatment with 100 μg/mL and 200 μg/mL LEO-Tw reduced the viability to 23.7 ± 3.9% and 9.2 ± 0.8% vs. 200 μg/mL citral (28.43 ± 6.65%) and vs. control (100%) (Figure 1B). Compared to A375, in the other melanoma cell line, RPMI-7951, citral, LEO, and LEO formulations decreased viability vs. control; however, it was to a lesser degree than in A375 as follows: 54.21 ± 7.13% (citral), 56.22 ± 3.11% (LEO 100 μg/mL), 46.42 ± 5.05% (LEO 200 μg/mL), 42.91 ± 5.51% (LEO-PEG 100 μg/mL), 35.13 ± 5.41% (LEO-PEG 200 μg/mL), 40.88 ± 4.90% (LEO-Tw 100 μg/mL), and 31.17 ± 5.38% (LEO-Tw 200 μg/mL) (Figure 1C). Similarly, HT-29 colorectal cancer cells were affected in a dose-dependent manner: 40.6 ± 5.1% and 27.8 ± 6.4% of the cells remained viable after 100 μg/mL and 200 μg/mL LEO treatment, respectively; the viability decreased to 43.7 ± 3.9% and 16.4 ± 7.1% after 100 μg/mL and 200 μg/mL LEO-PEG treatment and to 28.2 ± 6.0% and 21.6 ± 6.4% after 100 μg/mL and 200 μg/mL LEO-Tw treatment. Citral decreased HT-29 cell viability to 46.24 ± 3.60% (Figure 1D). In MCF-7 cells, treatment with 200 μg/mL citral decreased cell viability to 46.20 ± 6.04%, whereas treatment with 100 μg/mL and 200 μg/mL LEO, LEO-PEG, and LEO-Tw decreased cell viability to 46.58 ± 6.90% and 36.52 ± 7.08% (LEO), 39.33 ± 6.38% and 22.72 ± 4.04% (LEO-PEG), and to 40.25 ± 8.91% and 25.57 ± 4.72% (LEO-Tw) (Figure 1E).

The morphology of HaCaT cells remained unaltered following treatment with LEO, LEO-PEG, and LEO-Tw, showing preserved cell shape, adherence, and increased confluence, as observed in Figure 2. In contrast, all tested compounds induced marked cytotoxic effects in the cancer cell lines A375, RPMI-7951, MCF-7, and HT-29; the morphological assessment revealed characteristic signs of cell death, including loss of adherence, cell shrinkage, and rounding (Figure 2).

2.4. LEO Increases ROS Production

The ROS production in A375, RPMI-7951, MCF-7, and HT-29 cell lines after 24 h treatment with LEO, LEO-PEG, and LEO-Tw (100 μg/mL) was measured using the 2′,7′–dichlorofluorescein diacetate (DCFDA) assay. *Tert*-butyl hydroperoxide (TBHP, 50 μM) was used as a positive control. The results show that in A375 cell lines, treatment with LEO alone and formulated with PEG-400 and Tween 20 can significantly increase ROS production vs. control (100) as follows: 162.2 ± 10.11 (LEO), 178.7 ± 9.61 (LEO-PEG), and 179.8 ± 14.87 (LEO-Tw) (Figure 3A). A similar increase in ROS production vs. control was also observed after the treatment with LEO, LEO-PEG, and LEO-Tw in RPMI-7951 (135.4 ± 17.25, 145.1 ± 13.73, and 143.2 ± 8.58), MCF- 7 (152.5 ± 16.55, 162.0 ± 14.94, and 159.7 ± 8.35), and HT-29 (127.6 ± 9.60, 134.9 ± 13.14, and 125.1 ± 11.51) cell lines (Figure 3B–D).

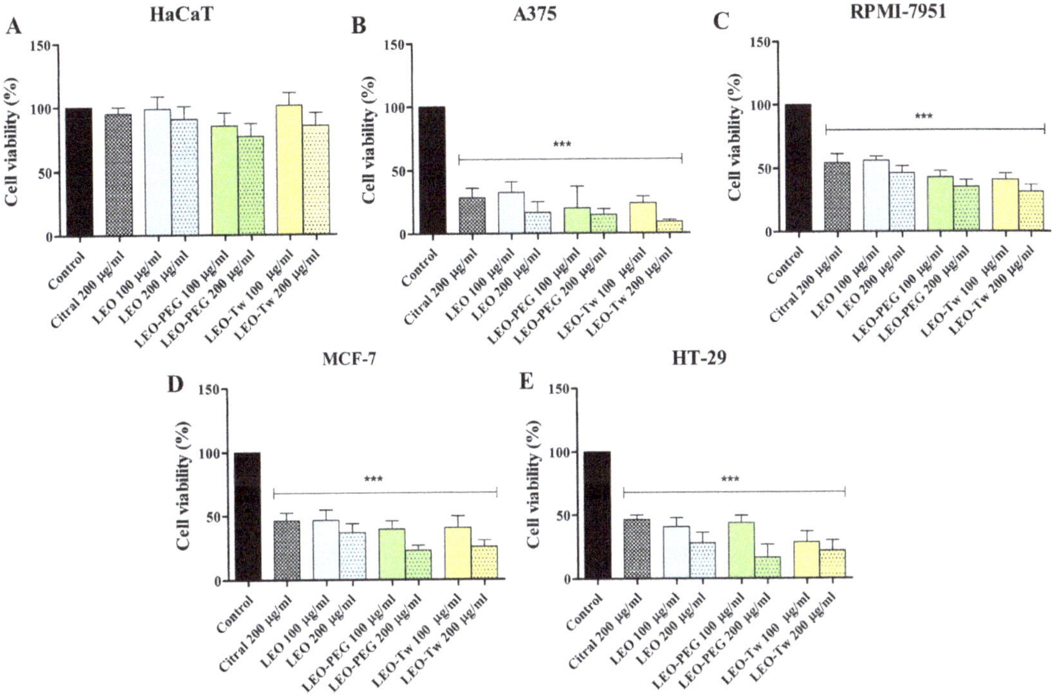

Figure 1. Cellular viability of HaCaT (**A**), A375 (**B**), RPMI-7951 (**C**), MCF-7 (**D**), and HT-29 (**E**) cells after 48 h stimulation with 100 µg/mL and 200 µg/mL simple LEO water dispersion, LEO-PEG, and LEO-Tw. The results are defined as cell viability percentage compared to control (100%) and expressed as mean values ± SD of three individual experiments (*** $p < 0.001$).

2.5. LEO Decreases Mitochondrial Membrane Potential

In healthy mitochondria, the monomer form (green fluorescence) of the JC-1 dye enters the mitochondria and forms JC-1 aggregates that emit red fluorescence; in contrast, in dysfunctional mitochondria where the normal MMP is disrupted (depolarization), the JC-1 cannot enter the mitochondria, and the aggregate/monomer ratio (red/green form) decreases. Thus, the changes in the MMP were evaluated by measuring the aggregate/monomer ratio in A375, RPMI-7951, MCF-7, and HT-29 cells exposed for 24 h to 100 µg/mL LEO, CEO-PEG, and CEO-Tw (Figure 4). Treatment with LEO, LEO-PEG, and LEO-Tw decreased the aggregate/monomer ratio in all tested cell lines, thus suggesting that LEO essential oil can induce MMP depolarization. Specifically, the strongest effect was observed in the A375 cell line as follows: 0.49 ± 0.08 (LEO), 0.47 ± 0.10 (LEO-PEG), and 0.42 ± 0.06 (LEO-Tw) (Figure 4A). In comparison, in RPMI-7951, the effects were less pronounced, with LEO decreasing the aggregate/monomer ratio to 0.79 ± 0.03, LEO-PEG to 0.69 ± 0.05, and LEO-Tw to 0.65 ± 0.07 (Figure 4B). In HT-29, the aggregate/monomer ratio decreased to 0.60 ± 0.05 (LEO), 0.54 ± 0.09 (LEO-PEG), and 0.56 ± 0.07 (LEO-Tw), whereas in MCF-7, the values were 0.52 ± 0.05 (LEO), 0.45 ± 0.09 (LEO-PEG), and 0.41 ± 0.06 (LEO-Tw) vs. control (1) (Figure 4C,D).

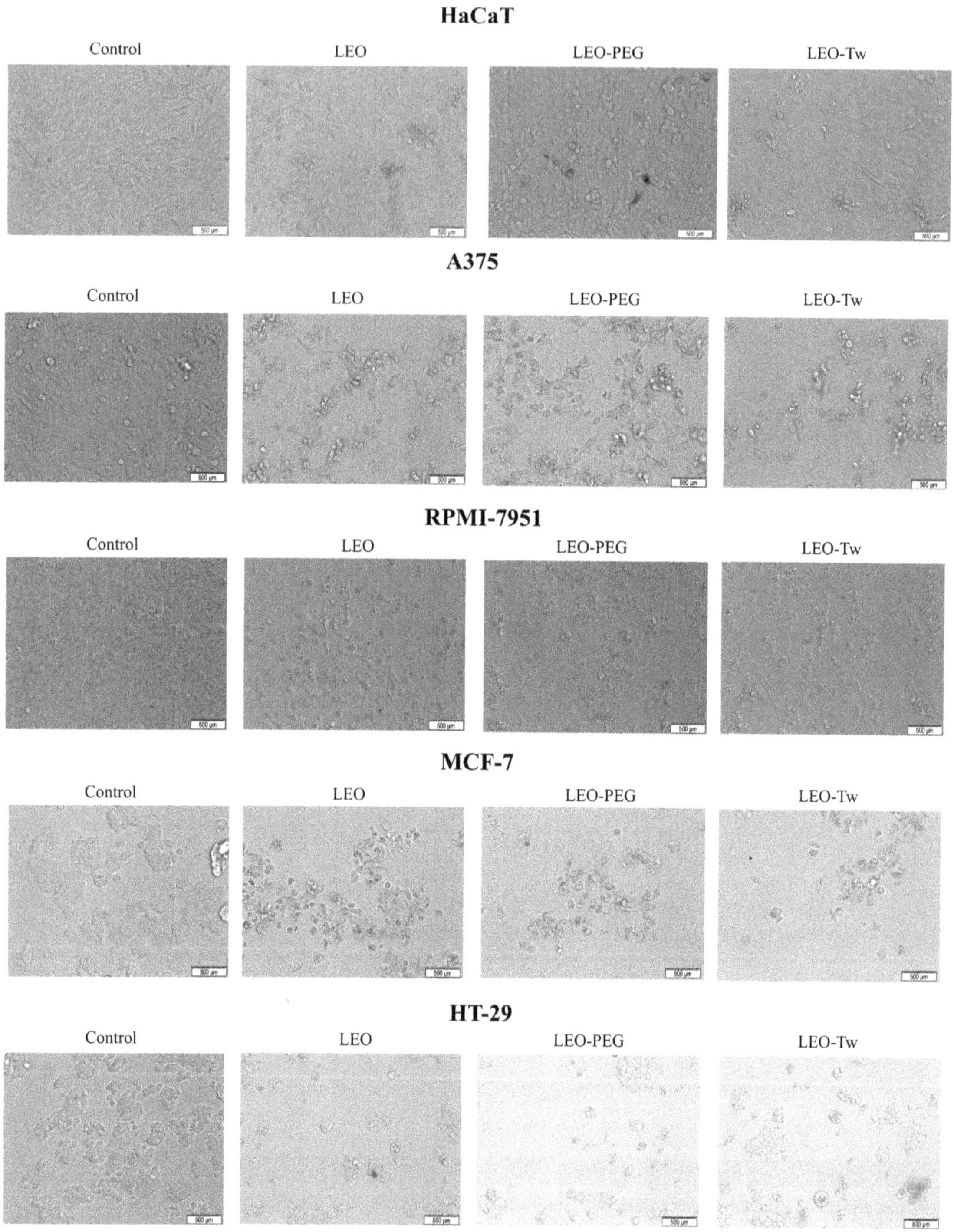

Figure 2. HaCaT, A375, RPMI-7951, MCF-7, and HT-29 cell line morphologies after 24 h treatment with LEO (1%), LEO-PEG, and LEO-Tw (200 μg/mL) vs. control.

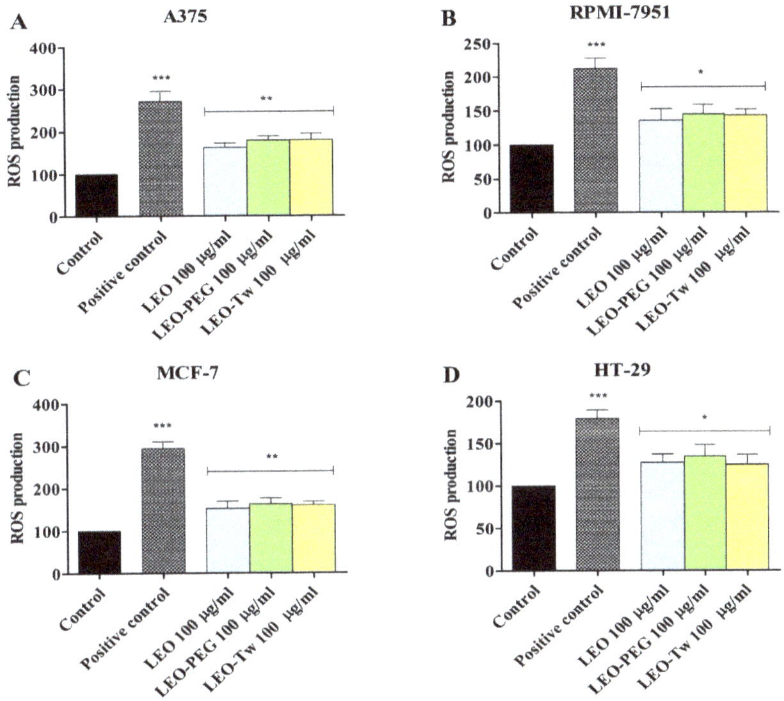

Figure 3. Assessment of ROS production in A375 (**A**), RPMI-7951 (**B**), MCF-7 (**C**), and HT-29 (**D**) cells exposed for 24 h to 100 µg/mL LEO, CEO-PEG, and CEO-Tw. *Tert*-butyl hydroperoxide (TBHP, 50 µM) was used as a positive control. The data represent the results of three independent experiments and are presented as the mean ± S.D (* $p < 0.05$, ** $p < 0.01$, and *** $p < 0.001$).

2.6. LEO Effect on Mitochondrial Function

The mitochondrial respiratory rates of permeabilized A375, RPMI-7951, MCF-7, and HT-29 cancer cells after treatment with 1% LEO was assessed using high-resolution respirometry. In all tested cancer cell lines, LEO significantly decreased all oxygen consumption rates (Figure 5, Table 3).

2.7. Molecular Docking of LEO Components

In the current study, a molecular docking-based method was employed to determine possible cancer-related protein targets for the 17 LEO components. The theoretical inhibition of these protein targets by LEO components could be related to the LEO in vitro anticancer cytotoxic activity. Subsequently, we docked the aforementioned compounds into the binding site of druggable protein targets whose overexpression is often correlated with carcinogenesis, increased cell proliferation, and survivability within various types of cancer; the chosen protein targets for the in silico experiment were the vascular endothelial growth factor receptor 2 (VEGFR2), the epidermal growth factor receptor 1 (EGFR1), dual specificity mitogen-activated protein kinase 1 (MEK1), phosphoinositide-dependent kinase-1 (PDK1), phosphatidylinositol 4,5-bisphosphate 3-kinase catalytic subunit alpha isoform (PI3Kα), phosphatidylinositol 4,5-bisphosphate 3-kinase catalytic subunit gamma isoform (PI3K γ), mammalian target of rapamycin (mTOR), protein kinase B (AKT/PKB), apoptosis regulator Bcl-2 (Bcl-2), and apoptosis regulator Bcl-XL (Bcl-XL). The obtained docking scores for compounds 1–17 and the native ligands (NLs) for each target, used as positive controls, are given in Table 4.

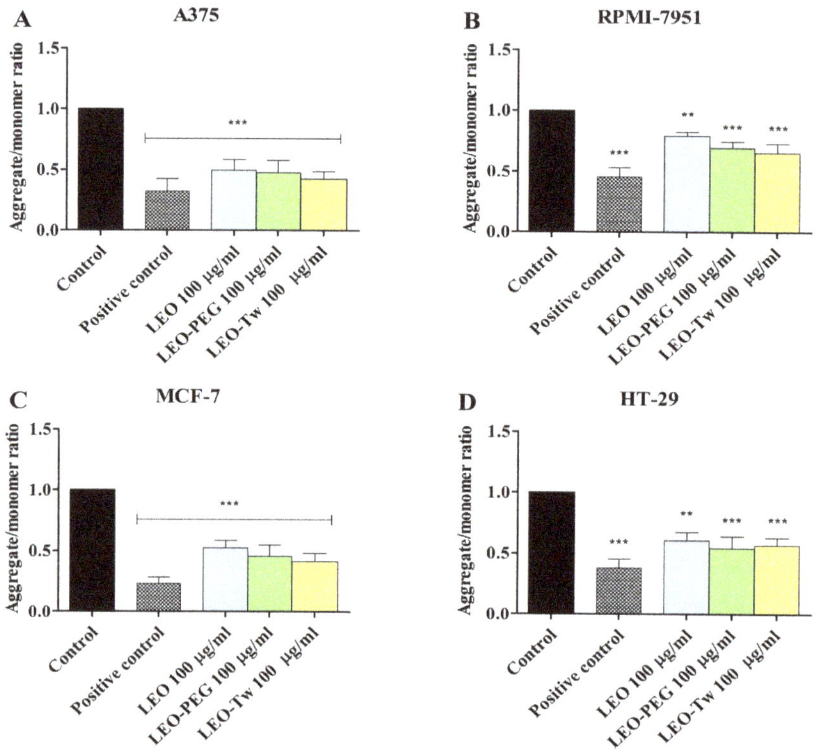

Figure 4. MMP, expressed as the JC-1 aggregate/monomer ratio of A375 (**A**), RPMI-7951 (**B**), MCF-7 (**C**), and HT-29 (**D**) cells exposed for 24 h to 100 μg/mL LEO, CEO-PEG, and CEO-Tw. FCCP (50 μM) was used as a positive control. The results are expressed as mean ± SD; n = 3 per group, ** $p < 0.01$, and *** $p < 0.001$.

Docking scores were recorded as binding affinity values (kcal/mol), which implies that the lower the negative value, the greater the inhibitory potential of that compound. Neither of the docked compounds showed lesser binding affinities than the NLs, which served as positive controls. Nonetheless, to compare the 17 compounds' combined effect against a specific protein target, all collected docking results were computed as a percentage of their respective NL's docking result; these values were plotted as a radar graph, with each corner representing one of the 10 protein targets used. If an overall inhibitory tendency towards certain proteins is present, the graph should show lines (indicating affinity values) that are orientated closer to one or more corners (proteins) of the graph. In this situation, the compounds' affinity lines are drawn closer to PDK1 and PI3Kα (Figure 6A). We divided the first graph into two subgraphs representing the major (compounds 2, 11, 12, 14, and 15) and minor components to see if the affinity trend holds in both situations. Figure 6B,C shows that there is a consistent preference for PDK1 and PI3Kα among LEO components. Having said that, among the major constituents, only compound 12 (β-caryophyllene) has a fairly substantial inhibitory potential for PDK1, scoring 85% of the native ligand's docking score. However, the cumulative contribution towards PDK1 and PI3Kα is, to a higher degree, attributed to the effect of minor components such as compound 17 (β-caryophyllene oxide), 13 (humulene), or 7 (α-cubebene). β-caryophyllene oxide (BCO) was the most active theoretical inhibitor of both PDK1 and PI3Kα, excluding the NLs. This aspect is clearly visible from the graphs, where the line corresponding to compound 17 is depicted in dark red (Figure 6B,C). Moreover, apart from the case of PI3Kγ, Bcl-XL, and mTOR, the same structure was overall ranked as the highest theoretically active compound.

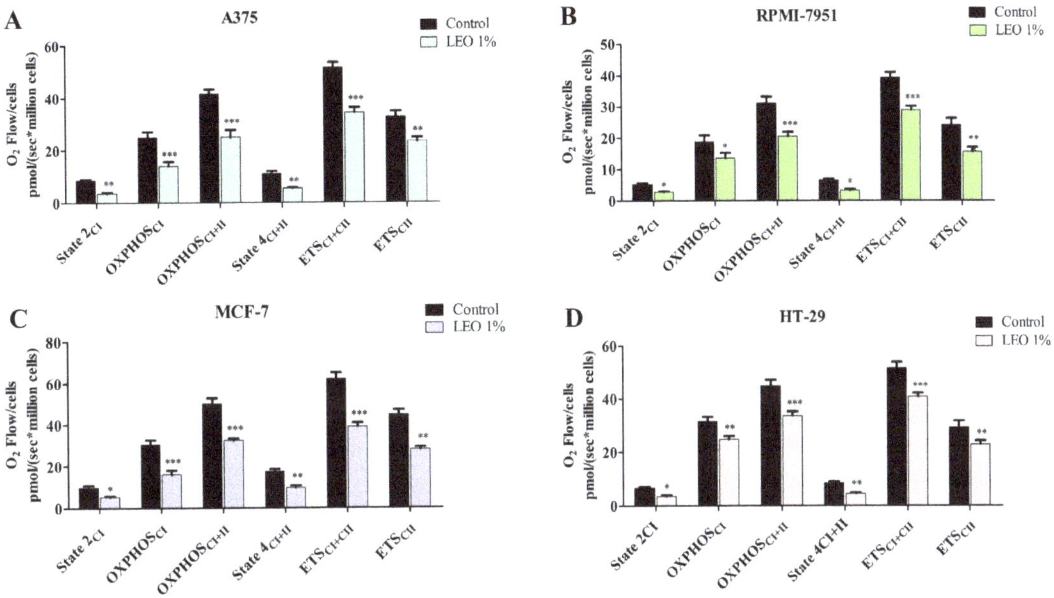

Figure 5. Mitochondrial respiratory rates of permeabilized A375 (**A**), RPMI-7951 (**B**), MCF-7 (**C**), and HT-29 (**D**) cells after treatment with 1% LEO. Results are expressed as mean values ± SD of three independent experiments (* $p < 0.05$, ** $p < 0.01$, and *** $p < 0.001$).

Table 3. The mean mitochondrial respiratory rates [pmol/(sec × million cells)] after the treatment of RPMI-7951, HT-29, A431, and NCI-H460 cells with eugenol and LEO.

		State 2	OXPHOS$_{CI}$	OXPHOS$_{CI+II}$	State 4	ETS$_{CI+II}$	ETS$_{CI}$
A375	Control	8.58	24.9	41.58	10.97	51.53	32.84
	LEO	3.64 **	13.97 *	25.03 ***	5.443 **	34.36 ***	23.32 **
RPMI-7951	Control	5.21	18.75	31.12	6.505	39.18	24.04
	LEO	2.84 *	13.56 ***	20.57 ***	3.31 *	28.79 ***	15.54 **
MCF-7	Control	9.69	30.45	50.02	17.49	62.03	44.67
	LEO	5.32 *	16.07 ***	32.53 ***	9.66 **	38.88 ***	27.89 **
HT-29	Control	6.603	31.56	44.90	8.40	51.60	29.32
	LEO	3.57 *	24.88 **	33.62 ***	4.35 **	40.72 ***	22.91 **

(* $p < 0.05$, ** $p < 0.01$, and *** $p < 0.001$).

In the case of PDK1, β-caryophyllene oxide interacts with the binding site of the target protein exclusively through hydrophobic interactions with amino acids such as LEU88, VAL96, ALA109, LEU159, and LEU212 (Figure 7). The same compound exhibits a similar binding pattern when docked into the PI3Kα binding site. The molecule establishes multiple hydrophobic interactions with adjacent amino acids (TRP780, ILE800, TRP836, VAL850, VAL851, MET922, PHE930, ILE932), but it also interacts with the binding site via a hydrogen bond with VAL851. The interaction binding pattern is depicted in Figure 8.

Table 4. Recorded docking scores (binding affinity) for the 17 LEO components.

Compound	Protein Targets									
	Egfr1	Vegfr2	Mek1	Akt	PI3Kα	mTOR	Bcl-2	Bcl-XL	PI3Kγ	PDK1
	Binding Affinity (kcal/mol)									
NL	−10.9	−12.1	−9.4	−9.4	−8.8	−11.2	−11.3	−10.8	−9.3	−8.7
1	−6	−7	−6.3	−6.3	−5.8	−6.3	−5.9	−5.9	−5.9	−6.5
2	−5.9	−6.5	−6.6	−6	−5.7	−6	−5.5	−6	−6	−6.2
3	−5.9	−5.9	−5.4	−5	−5.1	−4	−5.5	−6.2	−5.2	−5
4	−5.7	−6.6	−6.6	−5.9	−5.7	−5.9	−5.8	−5.9	−6.1	−6.2
5	−5.1	−5.7	−5.2	−4.9	−5.1	−4.9	−4.8	−5.2	−5.2	−4.8
6	−5.5	−6.1	−5.4	−5	−5.6	−5.5	−5.1	−5.4	−5.9	−5.6
7	−7.1	−6.5	−6.4	−6.3	−6.5	−5.9	−6.6	−7.8	−6.9	−7.3
8	−5.6	−6.1	−5.8	−5.9	−5.6	−5.7	−5.1	−5.5	−5.4	−5.7
9	−7	−6.9	−6.7	−5.8	−6.8	−5.8	−6.8	−6.9	−6.8	−6.4
10	−5.1	−5.6	−5.5	−5.3	−4.9	−5	−4.9	−5.2	−5.1	−5
11	−6.3	−6.8	−5.7	−5.2	−5.9	−5.5	−5.6	−5.7	−6.4	−5.6
12	−7.2	−7.1	−6.6	−6.2	−6.6	−6.8	−6.9	−7	−6.5	−7.4
13	−6.9	−6.7	−6.7	−5.8	−6.3	−5.7	−6.8	−7.2	−6.2	−7.5
14	−5.5	−6.2	−5.5	−5.5	−5.9	−5.3	−5.2	−5.6	−6	−5.6
15	−5.5	−6.3	−5.4	−5.3	−5.7	−5.8	−5.5	−5.5	−6.2	−5.4
16	−6.8	−7	−6.1	−5.9	−6	−6	−5.3	−5.9	−6.5	−6
17	−7.1	−7.3	−6.8	−6.4	−6.5	−6.3	−7.1	−7	−6.5	−7.9

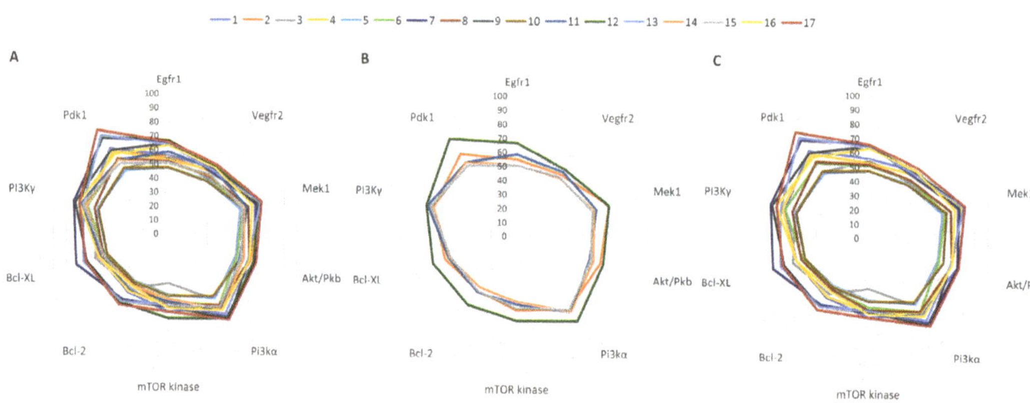

Figure 6. Radar graphs illustrating the docking scores of all docked components of LEO (**A**), its major constituents (**B**), and its minor components (**C**). The graph lines represent percentage values of each compound's docking score relative to their respective NL docking score.

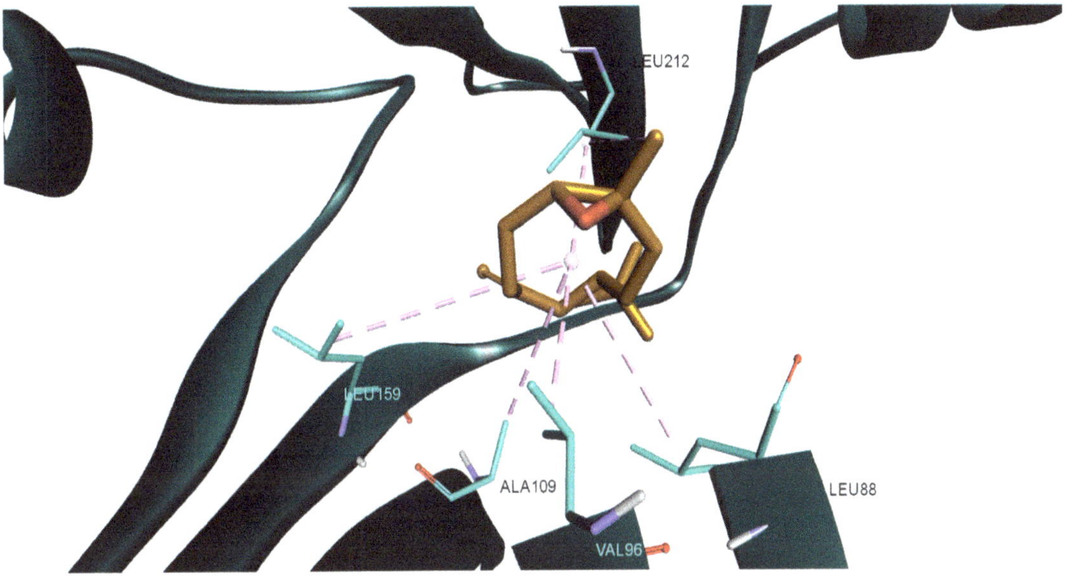

Figure 7. Compound 17 (β-caryophyllene oxide) docked into the binding site of PDK1 (PDB ID: 2PE1); hydrophobic interactions are depicted as purple dotted lines.

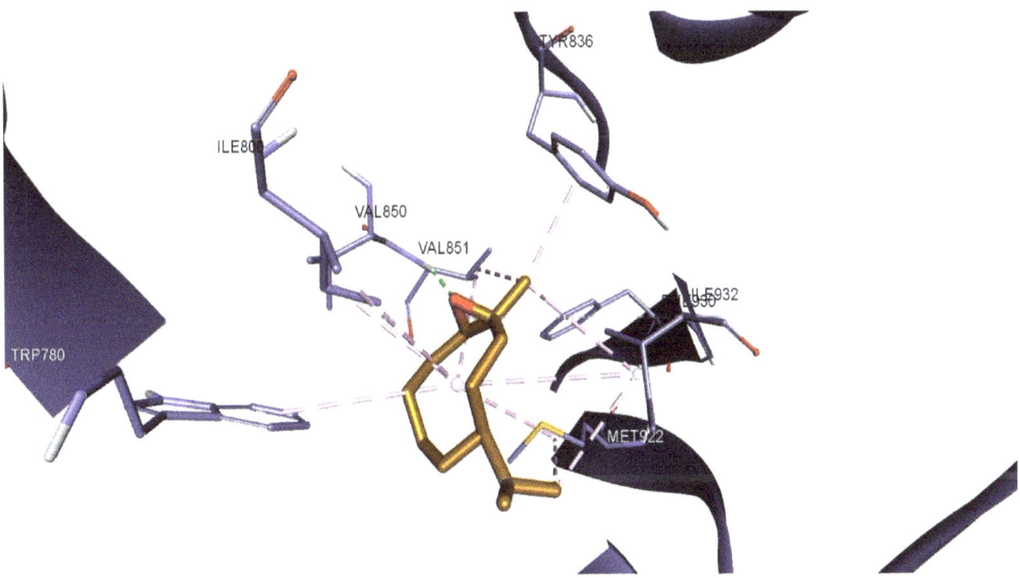

Figure 8. Compound 17 (β-caryophyllene oxide) docked into the binding site of PI3K α catalytic subunit (PDB ID: 6GVF); hydrophobic interactions are depicted as purple dotted lines, while hydrogen bonds are depicted as green dotted lines.

3. Discussion

Cancer represents one of the main causes of mortality worldwide. Although 36 types of this pathology have been identified, the majority of cancer-related deaths are being caused by melanoma, breast, and colorectal cancer [12–14]. Melanoma represents the deadliest form of skin cancer due to metastasis formation and rapid development of drug resistance [12,15]. Similarly, the treatment of breast and colorectal cancer is often inefficient due to the appearance of chemoresistance and immunoresistance [16–18]. As a result, there is a search for new and advanced methods of treatment that would surpass the problem of drug resistance. In this context, medicinal plants have come forth as a new promising alternative [19]. In recent years, there have been several in vitro and in silico studies of plants and their products regarding anticancer effects [20]. Lemongrass emerged as a potential source of antitumor compounds due to its essential oil content, with 1–10% in plant floral tops and leaves, thus making lemongrass superior to other plants in terms of essential oil quantity [21,22]. The extraction method was chosen to suitably meet our purpose and infrastructure needs. The composition of the obtained essential oil was highly similar to other ones reported in terms of citral composition; other less occurring oil components may vary depending on several variables [21,23,24]. Even if other extraction methods such as microwave-assisted hydrodistillation (MAHD) or supercritical fluid extraction (SED) produce higher oil yields, the citral content does improve significantly and is dependent on other extraction factors [25]. The same study suggests that our chosen extraction method is the least expensive and environmentally harmful method for extracting lemongrass oil on a large scale. MAHD is appropriate for smaller operations, since it offers energy economy and shorter extraction times. Although SFE produces high-quality oil, it is more expensive and only works well in specific applications [25].

On the same note, DPPH scavenging activity of the LEO was not superior to the ones recorded for ascorbic acid or BHA but is in the same range as other values reported and can be mostly attributed to the high citral concentration [23,24]. LEO showed very close ABTS radical inhibition to both controls at a concentration of 1 mg/mL. This implies that LEO might be more efficient in scavenging ABTS radicals than DPPH radicals, possibly due to their difference in solubility and reactivity. A previous study showed that the major component, cital, showed higher free radical scavenging ability in the ABTS•+ assay than in the DPPH• assay [26]. Thus, citral's antioxidant activity is more effectively determined by the ABTS method and is related to the same behavior exhibited by the LEO in the same scenario. The use of LEO could bring benefits to cancer therapies because of its biocompatibility, natural origin, and low cost [21]. Nevertheless, one of the disadvantages that essential oils have is their low solubility in water [22]. Therefore, in order to increase the homogeneity of essential oil–culture media mixture prepared in this study, we used Tween 20 and polyethylene glycol (PEG) 400. Tween 20 is a non-ionic surfactant with a HLB value of 16.7, thus forming emulsions with a suitable viscosity for drug formulations [27]. Also, there are multiple cytotoxic studies where Tween 20 was used for essential oil emulsification [27,28]. Likewise, PEGs are biocompatible non-toxic polymers often used as cosurfactants in pharmaceutical applications, PEG-400 being the most used subtype in self-emulsifying drug delivery systems [29–32]. After cell stimulation, it was observed that Tween and PEG-400 do not have any toxic effects on healthy and malignant cells, confirming their suitability for cytotoxic assays.

One of the main problems regarding cancer therapy is the poor selectivity for cancerous cells, with many side effects being caused by the toxic effects of the drug on healthy cells. Therefore, researchers aim to uncover highly selective molecules with anticancer properties [33]. In our case, neither one of the tested concentrations of LEO, LEO-PG, or LEO-TW significantly affected the viability of healthy keratinocytes, thus showing promising selectivity characteristics. Similar results were obtained by Al-Ghanayem et al., who reported that the cell viability of HaCaT cells was above 90% after treatment with *Cymbopogon flexuosus* essential oil (160 μg/mL, in culture media), the IC_{50} value being 1250 μg/mL [34]. On the contrary, the concentrations of LEO (0.5% and 1% *v*/*v*) that did not affect HaCaT cells showed toxic effects when tested on human dermal fibroblasts, the activity being dose-dependent [35]. This could be explained by the fact that different cell lines present different susceptibilities to tested compounds; also, the total concentration of citral (>89%) in the essential oil tested on fibroblasts was higher compared to our study (73.5%) [8,35].

On the other hand, all tested concentrations of LEO-CM, LEO-PEG, and LEO-TW presented cytotoxic effects on the colorectal adenocarcinoma, breast cancer, and melanoma cell lines. It was observed that higher concentrations (1% and 200 μg/mL) exhibited a stronger toxic effect compared to lower concentrations (0.5% and 100 μg/mL), with the anticancer activity being dose-dependent. A similar study was conducted by Sharma et al., who tested the effects of a LEO that was different in terms of chemical composition (*Cymbopogon flexuosus*) on HT-29 cells and other colon cancer cells (HCT-15, SW-620, and 502713). In agreement with our results, the study showed that lemongrass essential oil has a toxic effect on this type of cancer, with IC_{50} values of 42.4 μg/mL, 60.2 μg/mL, 28.1 μg/mL, and 4.2 μg/mL, respectively [7]. Whilst both oils shared some common components (linalool, citronella, and caryophyllene oxide), the major constituents in Sharma's case were isointermedeol (24.97%), geraniol (20.08%), and geranyl acetate (12.20%) [36], while in our case, α-Citral (37.44%), β-Citral (36.06%), and linalool acetate (9.82%) were identified; despite their different compositions, both essential oils showed cytotoxic activity supporting the notion that multiple constituents have effects on cancerous cells [21,25]. Similar results were also described by Wang et al. [37], who revealed that Lemon myrtle (*Backhousia citriodora* F.Muell.) essential oil, rich in citral, (accounted for 74.9% of the content) had an IC_{50} value of 68.91 ± 4.62 on the HT-29 cell line. In a more recent study, the cytotoxic effect of LEO on HT-29 cells was attributed to citral and geraniol, compounds that were able to induce intrinsic apoptosis in malignant cells while leaving healthy cells unharmed [38]. α-citral, also known as geranial, and β-citral, known as neral, are aldehyde monoterpenes that form a racemic mixture (citral) which is considered an indicator of essential oil quality [21]. In recent years, there have been many studies that have focused on citral's anticancer activity [39]. For example, Sheikh et al. revealed that citral has a dose-dependent toxicity on HT-29 (IC_{50} was 181.21 μM) and HCT116 (IC_{50} was 145.32 μM) after 24 h stimulation; moreover, the authors reported that cell death was induced through increased ROS levels and p53 activation caused by citral [13].

On melanoma cells, another study showed that citral has a cytotoxic effect, with IC_{50} values of 11.7 μM (SK-MEL-147) and 13.4 μM (UACC-257). The UACC-257 cell line, as well as the A375 line used in our study, harbors the BRAF mutation; the authors indicated that citral could reduce the effects of this mutation by interfering with the MAPK pathway and therefore interfere with carcinogenesis [15]. Aside from citral, other major components found in the LEO were researched for their anticancer potential. Several studies reported that linalool presents cytotoxic effects, the tested cell lines being those of colon cancer (WiDr, HCT-15, SW480 and RKO), lung cancer (A549), and ovarian cancer (HeyA8, A2780, and SKOV3ip1) [40]. While many studies have reported citral's and linalool's cytotoxic effects, it was often observed that essential oils have stronger activity compared to their individual constituents, possibly due to the synergistic relationship between the associated compounds [21]. These reports consolidate our findings, where the concentration of 200 μg/mL citral was less effective compared to the same concentration of LEO in all investigated scenarios. Research that confirms this observation was made by Gaonkar et al., who studied the effects of both *Cymbopogon flexuosus* essential oil and its components, citral and geraniol, in MCF-7 and HEK-293 cells. The authors also reported that the cell lines presented variant levels of sensitivity to tested compounds, with the MCF-7 cell line being more sensitive [8]. It is, therefore, safe to assume that anticancer activity is exerted through several molecular mechanisms shared by two or more individual components, with the association of complementary pathways being able to converge towards a synergistically improved overall anticancer outcome.

We investigated the underlying mechanisms of the LEO's biological effects and identified its stimulatory activity on ROS production in all tested cancer cells. The induction of ROS by lemongrass oil was previously hypothesized by Lee et al., who assessed its antifungal activity [41]; in fact, its main component, citral, exhibited a similar ROS generation effect which was indicated as responsible for the significant anti-melanoma effects [15]. However, the intimate mechanism is more complex—on one hand, all stages in melanoma development involve oxidative stress and on the other, citral exerts cytotoxic effects through the induction of oxidative stress, as revealed by ROS generation. It is therefore safe to assume that the cytotoxic activity implies not only the generation of ROS but also a modulation of the intracellular pathways related to DNA damage, cellular proliferation, and death. In MCF-7 cells, our study revealed a stronger inhibitory effect on cell viability compared to previous data [42]; however, a more recent study reported significant anticancer activity against MCF-7 cells through mitochondrial depolarization and ROS generation [43]. Our results are in agreement with such studies, since we found a strong mitochondrial membrane depolarization in the A375, MCF-7, and HT-29 cells and at a lower degree in RPMI-7951, which is indicated as an important step of apoptosis. Changes in the mitochondrial morphology as a result of ROS production were previously noted in a different cell line of colon cancer, SW1417, where LEO induced mitochondrial fission in a dose-dependent manner, finally leading to apoptosis [44].

The increase in ROS production after LEO treatment can be further understood by evaluating the mitochondrial function, as the mitochondria are the primary site of ROS generation, specifically at the level of the complex I (CI) and complex III (CIII) of the electron transport chain (ETC) [45]. Increased ROS production in cancer cells has been associated with mitochondrial dysfunction, and the disruption of the mitochondrial ETC has been shown to lead to an increased ROS production in various cancer cell lines [46–49]. Our findings are similar; treatment with LEO increased ROS production and, at the same time, decreased all mitochondrial respiratory rates, suggesting that LEO can induce the impairment of mitochondrial function/mitochondrial dysfunction. Indeed, other studies have demonstrated that citral, the major component of LEO, has the ability to induce mitochondrial dysfunction by disrupting the tricarboxylic acid cycle (TCA) pathway and by damaging the mitochondrial membrane permeability of *Penicillium digitatum* cells [50]. Another study revealed that on the same P. digitatum cells, citral drastically affects all the mitochondrial complexes involved in oxidative phosphorylation, namely CI-CV. Moreover, citral significantly decreased intracellular ATP and the mitochondrial membrane MMP while increasing the accumulation of ROS [51]. In a similar manner, the current study showed that LEO induced mitochondrial dysfunction, increased ROS production beyond the threshold that cancer cells can tolerate, and decreased the MMP, ultimately leading to cell death. This dual nature, both promoting ROS production and being associated with antioxidant activity, is not unique to LEO and appears to be dependent on the cellular context and redox state. As revealed in the review by Bezzera et al. and in one of our recent studies, eugenol, the main component of clove essential oil, has similar antioxidant and pro-oxidant activities, acting as antioxidant in normal cells while displaying pro-oxidant cytotoxic effects in cancer cells depending on the cellular redox environment [52,53]. As previously mentioned, our experimental findings indicate that LEO exhibited superior cytotoxicity against cancer cells compared to citral. We also employed an in silico method to examine if the anticancer impact of LEO might be partially ascribed to the hypothesized suppression of active target proteins that promote carcinogenesis when overexpressed. Our docking results showed that among all the targets tested, PDK1 and PI3Kα were the preferred targets by the majority of LEO components. Minor components, rather than both citral isomers, were primarily responsible for the inhibition of these proteins. The strongest inhibitory potential for both proteins was attributed to CPO. Sadly, there are no experimental data regarding the biological inhibition of PDK1 by CPO. However, there is evidence that this compound targets PI3K. The results of Park et al. showed that CPO causes apoptosis by increasing mitochondrial ROS production in addition to inhibiting the constitutive activation of the PI3K/AKT/mTOR/S6K1 signaling pathway in PC-3 human prostate and MCF-7 breast cancer cells [54]. The inhibition activity of CPO may be closely related to its showcased binding pattern, which is similar to that of other PI3Kα inhibitors. Sapanisertib and a few other PI3Kα inhibitors, described in the study by Ouvry et al., form an essential hydrogen bond with the hinge region amino acid VAL851 [55]. The same interaction is present in the CPO-PI3Kα docked complex, which contributes to the compound's high inhibition score among all other LEO constituents.

Thus, all presented findings indicate that LEO's cytotoxicity could be attributed to the combined activity of its components and multiple modes of action. Effects on melanoma and colorectal cancer cells, along with good selectivity and possible reduction in drug resistance, all show that lemongrass essential oil has a potential as an anticancerous agent.

4. Materials and Methods

4.1. LEO Extraction and GC-MS Analysis

The dried plant material was received as a gift from "King Michael I" University of Life Sciences (Timisoara Romania Herbarium, voucher number VSNH.BUASTM-128). The dried plant was grounded and then was subjected to steam hydrodistillation for 4 h at 100 °C by means of a Craveiro-type apparatus [56,57]. As previously described [58], the steam was generated by heating a 3000 mL glass boiler equipped with electrical resistance and previously filled with water. The boiler was refilled whenever necessary. The steam was then transferred to the bottom of the glass extraction vessel (1000 mL). A water-cooling system was used to condense the steam and vaporized oil after it had passed through the plant material, which was placed on a perforated plate a few centimeters from the base of the extraction tank. Lastly, to prevent the creation of artifacts from overheating, LEO and hydrosol (aqueous phase) were collected in a 250 mL glass receiver fitted with a water-cooling jacket and a hydrosol overflow outlet. Following separation, the oil was treated with anhydrous sodium sulfate in order to remove water traces and stored at −18 °C in sealed vials for future analysis. The extraction yield was calculated according to the following formula: oil weight/dried plant weight × 100 (wt).

The sample was analyzed by gas chromatography using a GC Hewlett Packard HP 6890 Series gas chromatograph coupled with a Hewlett Packard 5973 Mass Selective Detector. Briefly, 1 μL of diluted sample (1:100 in hexane) was injected into the gas chromatograph under the following parameters: DB-WAX capillary column (30 m length, 0.25 mm internal diameter, 0.25 μm film thickness), a 50 °C to 250 °C temperature range with a rate of 6 °C/minute, and 4 min solvent delay. The mass spectrometer was set to 230 °C with the MS Quad at 150 °C and helium gas flow at 1 mL/min. The analyzed compounds ranged in mass from 50 to 600 amu. The resulting spectra were assessed against data from NIST 02 library (USA National Institute of Science and Technology software), and area percentage was determined. The retention indexes were calculated based on the retention times and areas of C9 to C18 alkanes; also, the Adams Indexes were used for comparation with literature.

4.2. DPPH Antioxidant Scavenging Activity Determination

The antioxidant activity of the samples was evaluated using the 2,2-diphenyl-1-picrylhydrazyl (DPPH) radical scavenging assay, as described by Rădulescu et al. [59]. A DPPH stock solution was prepared by dissolving 5 mg of DPPH in 5 mL of ethanol. Serial dilutions were then made in ethanol to generate a calibration curve with concentrations ranging from 7.81 μg/mL to 0.5 mg/mL. For comparative analysis, positive controls, ascorbic acid (AA), and butylated hydroxyanisole (BHA) were prepared in a concentration range between 0.06 μg/mL and 1.2 mg/mL.

The plant extract was diluted 1:10 in ethanol and mixed with a 0.25 mM DPPH ethanolic solution in a 1:4 (*v*/*v*) ratio. The reaction mixture was incubated in the dark at 25 °C for 30 min. Absorbance was measured at 515 nm using a Tecan Infinite 200Pro spectrophotometer (Tecan Group Ltd., Männedorf, Switzerland) with i-control software (version 1.10.4.0).

The percentage of DPPH radical inhibition (Inh%) was calculated using the following formula:

$$\text{Inh\%} = [(A_0 - A_s)/A_0] \times 100$$

where A_0 is the absorbance of the control (DPPH solution without the sample) and A_s is the absorbance of the sample. The inhibition values were plotted against sample concentrations to determine the IC_{50} (half-maximal inhibitory concentration) using the calibration curve equation specific to each sample and control. The IC_{50} values, expressed in μg/mL, provide a quantitative measure of the sample's antioxidant capacity, with lower values indicating higher antioxidant activity.

4.3. ABTS Radical Scavenging Assay

The ABTS [2,2′-azino-bis(3-ethylbenzthiazoline-6-sulfonic acid)] radical scavenging activity was assessed using a modified method [58,60]. To generate the ABTS cation (ABTS+•), a solution of ABTS (7.29 mM) was mixed in a 1:1 (*v*/*v*) ratio with $K_2S_2O_8$ (2.47 mM) in an amber-colored bottle and kept in the dark at 25 °C for 14 h. The resulting ABTS+• solution was then diluted with ethanol to achieve an absorbance of 0.745 ± 0.047 at 734 nm. Next, 400 μL of the ABTS+• solution was combined with 100 μL of 1:10 diluted sample extract and incubated at room temperature in the dark for 30 min. The absorbance was subsequently measured at 734 nm using a 1 mg/mL solution of ascorbic acid (AA) and butylated hydroxyanisole (BHA) as reference compounds. The following equation was used:

$$\%\text{ABTS+}\bullet\ \text{inhibition} = (A_{control} - A_{sample}) \times 100/A_{control},$$

where $A_{control}$ measures the absorbance of the ABTS+• solution mixture without adding the sample and A_{sample} measures the absorbance of the sample with ABTS+• solution mixture.

4.4. Cell Culture

Cell lines used for this study were immortalized human keratinocytes HaCaT, obtained from CLS Cell Lines Service GmbH (Eppelheim, Germany), and human melanoma cells A375 and human colorectal adenocarcinoma cells HT-29, both obtained from American Type Culture Collection (ATCC, Lomianki, Poland). HaCat and A375 cell lines were cultured in DMEM (Dulbecco's Modified Eagle Medium) High Glucose, with the addition of 10% FBS (fetal bovine serum) and a 1% penicillin/streptomycin mixture (100 IU/mL), whereas HT-29 was cultured in McCoy's 5A Medium, with 10% FBS and a 1% penicillin/streptomycin mixture (100 IU/mL). Cells were incubated in 5% CO_2 atmosphere at 37 °C.

4.5. Cell Viability Assessment

Cell viability was determined using the Alamar Blue assay. Cell lines were seeded in 96-well culture plates (10^4 cells/plate) and kept in an incubator at 37 °C and 5% CO_2. After reaching 80–90% confluence, the old media was removed and the cells were stimulated using three methods as follows: (1) water dispersion method: lemongrass essential oil was mixed with culture media (LEO) to obtain 100 and 200 μg/mL that were used for cell treatment; (2) PEG method: lemongrass essential oil was mixed with PEG-400 (LEO-PEG) to obtain a 10 mg/mL mixture that was sonicated and thereafter diluted with culture media to obtain a 1 mg/mL concentration, and the respective mixture was diluted once more in culture media to obtain the final concentrations of 100 μg/mL and 200 μg/mL used for cell treatment; and (3) the Tween method, where Tween 20 (LEO-Tw) was used instead of PEG-400, with the remainder of the method being identical to the others. After 48 h of incubation with the test compound, the old medium was removed from the wells and the new medium containing Alamar Blue was added (final concentration 1.5%). Cells were incubated in standard conditions for another 4 h. Subsequently, the absorbance was measured at 570 nm using the xMark™ Microplate Spectrophotometer, Bio-Rad. Citral analytical standard, the main component of LEO, was purchased from Merck (Merck KGaA, Darmstadt, Germany) and also tested in the same conditions mentioned above at a concentration of 200 μg/mL.

4.6. Assessment of Cellular ROS Production

ROS production was evaluated using the cell permeant 2′,7′-dichlorodihydrofluorescein diacetate (H2DCFDA) kit (ab113851, Abcam, Cambridge, UK). After 24 h of treatment with 100 μg/mL citral, LEO, LEO-PEG, and LEO-Tw, the ROS production in A375, RPMI-7951, MCF-7, and HT-29 cell lines was quantified at Ex/Em at 485/535 nm using the BioTek Synergy HTX multimode microplate reader (Agilent Technologies, Santa Clara, CA, USA), following the protocol described by the manufacturer [61].

4.7. JC-1 Assay

The mitochondrial membrane potential was assessed using the JC1 Kit (JC1- Mitochondrial Membrane Potential Assay Kit ab113850, Abcam, Cambridge, MA, USA), according to the manufacturer's specifications [62]. The JC-1 cationic dye exhibits a potential-dependent accumulation in mitochondria, as observed by the shift in fluorescence emission from green to red. The stained A375, RPMI-7951, MCF-7, and HT-29 cells treated with 100 μg/mL citral, LEO, LEO-PEG, and LEO-Tw were analyzed (Ex/Em: 535/590 nm) using a multimode microplate reader (BioTek Synergy HTX multimode microplate reader, Agilent Technologies, Santa Clara, CA, USA).

4.8. High-Resolution Respirometry

High-resolution respirometry was used to assess the mitochondrial function at 37 °C by means of the Oroboros high-resolution respirometer (Oxygraph-2k Oroboros Instruments GmbH, Innsbruck, Austria). A modified substrate uncoupler–inhibitor–titration (SUIT) protocol previously described by Petruș et al. was used to determine the mitochondrial respiratory rates [63]. A375, RPMI-7951, HT-29, and MCF7 cells were cultured until reaching confluence, trypsinized, counted, and resuspended (1×10^6/mL cells) in a special mitochondrial respiration medium (EGTA 0.5 mM, 3 mM KH2PO4, taurine 20 mM, K-lactobionate 60 mM, $MgCl_2$ 10 mM, D-sucrose 110 mM, HEPES 20 mM, and BSA 1 g/L, pH 7.1). The cells were added to the device chambers after 15 min in order to allow for the stabilization of the oxygen flux. In the first step, digitonin (20–35 μg/1×10^6 cells) was added to permeabilize the cell membrane, followed by glutamate (10 mM) and malate (5 mM) as the complex I (CI) substrates that allowed for the measurement of endogenous

ADP-dependent basal respiration, also known as State 2. In the second step, ADP (5 mM) was added in order to enable the measurement of the CI (OXPHOSCI)-dependent active respiration, followed by the addition of succinate (10 mM) as the complex II (CII) substrate, thus enabling the measurement of active respiration dependent on both CI and CII (OXPHOSCI + II). Complex V was then inhibited with oligomycin (1 μg/mL), which resulted in the measurement of CI- and CII-dependent leak respiration (State 4). P-(trifluoromethoxy) phenylhydrazone carbonyl cyanide—FCCP—was further titrated in steps (1 μM/step) in order to achieve and measure the maximal respiratory capacity of the electron transport system (ETSCI + II). A CI inhibitor, rotenone (0.5 μM,), was then added to obtain the ETS dependent solely on CI. In the final step, antimycin A (2.5 μM) was added as a CIII inhibitor for the purpose of inhibiting the mitochondrial respiration and allowing for the measurement of the residual oxygen consumption (ROX). The final values were recorded following ROX correction.

4.9. Molecular Docking

Molecular docking studies were conducted following a previously described workflow [52]. Briefly, the protein target structures were searched in the RCSB Protein Data Bank [64] and optimized as docking targets by using Autodock Tools v1.5.6 (The Scripps Research Institute, La Jolla, CA, USA) (Table 5). The SDF structure files corresponding to the essential oil components were retrieved from PubChem [65] and converted into 3D structures by means of PyRx's Open Babel module [66]. PyRx v0.8 [66] virtual screening software (The Scripps Research Institute, La Jolla, CA, USA) was employed in order to achieve molecular docking by using Vina's encoded scoring function [67]. The validation of the docking protocol was conducted by the re-docking of the native ligands into their original protein binding sites. The root mean square deviation (RMSD) between the predicted and experimental docking pose of the native ligand was calculated in order to use the determined values to identify cases where molecular docking should be performed (a RMSD value not exceeding 2 Å was used as a threshold). The docking grid box coordinates and size were selected to best fit the active binding site (Table 5). Docking scores were calculated as ΔG binding energy values (kcal/mol). Protein–ligand binding interactions were analyzed using Accelrys Discovery Studio Visualizer 4.1 (Dassault Systems BIOVIA, San Diego, CA, USA).

Table 5. Docking parameters used for the molecular docking of LEO components.

Protein (PDB ID)	Grid Box Center Coordinates	Grid Box Size	References
VEGFR2 (4ASD)	x = −23.4872 y = −1.3964 z = −11.0618	x = 14.6340 y = 19.2181 z = 15.7983	[68]
EGFR1 (1XKK)	x = 19.4479 y = 33.9295 z = 38.3514	x = 18.8523 y = 18.8523 z = 18.8523	[69]
MEK1 (3DV3)	x = 38.8352 y = −14.6371 z = 0.0462	x = 13.0549 y = 19.2181 z = 11.5471	[70]
PDK1 (2PE1)	x = −6.2833 y = 44.2844 z = 44.0516	x = 11.6743 y = 14.1528 z = 11.6743	[71]

Table 5. *Cont.*

Protein (PDB ID)	Grid Box Center Coordinates	Grid Box Size	References
AKT/PKB (4GV1)	x = −19.9894 y = 3.3152 z = 11.0426	x = 14.2711 y = 13.5948 z = 15.4926	[72]
PI3Kα (6GVF)	x = −17.2065 y = 147.5732 z = 29.1217	x = 14.8956 y = 14.8956 z = 14.8956	[55]
PI3Kγ (4FA6)	x = 44.2130 y = 13.1865 z = 29.7323	x = 13.5948 y = 13.5948 z = 13.5948	[73]
mTOR (4JSX)	x = 50.4459 y = −2.0684 z = −48.5963	x = 14.2711 y = 13.5948 z = 15.4926	[74]
BCL-XL (2YXJ)	x = −9.7398 y = −16.3876 z = 8.8381	x = 17.9927 y = 26.4869 z = 15.9145	[75]
BCL-2 (4LVT)	x = 7.6196 y = −3.0737 z = −10.3894	x = 16.5597 y = 26.5085 z = 20.6849	[76]

4.10. Statistical Analysis

In order to determine the statistical differences compared to the control, the cell viability, ROS, and LP assay results were analyzed statistically using a one-way ANOVA followed by the Dunnett post-test. The high-resolution respirometry study results were analyzed using a two-way ANOVA with Bonferroni's multiple comparison post-test. All the differences were considered to be statistically significant if $p < 0.05$ (* $p < 0.05$, ** $p < 0.01$, and *** $p < 0.001$).

5. Conclusions

The current study highlights the significant in vitro anticancer potential of LEO, exhibiting its cytotoxic efficiency against multiple cancer cell lines while maintaining its selective effect on non-cancerous cells. Further investigations showed that the cytotoxic effect of LEO was mediated through increased ROS production, mitochondrial membrane depolarization, and impaired mitochondrial function. Moreover, molecular docking analysis revealed that minor components, namely β-caryophyllene oxide, may play a significant role in targeting key proteins such as PDK1 and PI3Kα. These findings support that LEO's anticancer activity is a cumulative effect attributed to several components as opposed to the major component, citral. While LEO represents a promising candidate for further investigations as an alternative or complementary anticancer therapy, future work encompasses in vivo anticancer efficiency validation and formulation strategies to enhance its bioavailability and therapeutic efficacy.

Author Contributions: Conceptualization, T.M., D.M. and A.G.; methodology, A.M. and A.G.; software, A.M. and M.M.; validation, A.G., C.Ș. and A.M.; formal analysis, A.M., D.C., A.T.L.-G., M.-A.P. and C.J.; investigation, T.M., D.M., M.M., D.C., A.T.L.-G., M.-A.P. and C.J.; data curation, T.M., D.M. and M.M.; writing—original draft preparation, T.M., D.M. and A.M.; writing—review and editing, C.Ș. and A.G.; visualization, C.Ș.; supervision, A.G. All authors have read and agreed to the published version of the manuscript.

Funding: We would like to acknowledge the "Victor Babes" University of Medicine and Pharmacy Timisoara for their support in covering the costs of publication for this research paper.

Institutional Review Board Statement: Not applicable.

Informed Consent Statement: Not applicable.

Data Availability Statement: The original contributions presented in the study are included in the article; further inquiries can be directed to the corresponding author.

Conflicts of Interest: The authors declare no conflicts of interest.

References

1. Duan, C.; Yu, M.; Xu, J.; Li, B.-Y.; Zhao, Y.; Kankala, R.K. Overcoming Cancer Multi-drug Resistance (MDR): Reasons, mechanisms, nanotherapeutic solutions, and challenges. *Biomed. Pharmacother.* **2023**, *162*, 114643. [CrossRef] [PubMed]
2. Chehelgerdi, M.; Chehelgerdi, M.; Allela, O.Q.B.; Pecho, R.D.C.; Jayasankar, N.; Rao, D.P.; Thamaraikani, T.; Vasanthan, M.; Viktor, P.; Lakshmaiya, N.; et al. Progressing nanotechnology to improve targeted cancer treatment: Overcoming hurdles in its clinical implementation. *Mol. Cancer* **2023**, *22*, 169. [CrossRef] [PubMed]
3. Mohamed, K.; Abarikwu, S.O.; Mmema, L.; Jibril, A.T.; Rahmah, L.; Ivanovska, M.; Rahimi, A.M.; Joya, M.; Hashem, F.; Essouma, M.; et al. A Global Perspective of Cancer Prevalence: The Causative Agent, the Environment, or the Genes? In *Handbook of Cancer and Immunology*; Springer International Publishing: Cham, Switzerland, 2023; pp. 1–21.
4. Lin, S.; Chang, C.; Hsu, C.; Tsai, M.; Cheng, H.; Leong, M.K.; Sung, P.; Chen, J.; Weng, C. Natural compounds as potential adjuvants to cancer therapy: Preclinical evidence. *Br. J. Pharmacol.* **2020**, *177*, 1409–1423. [CrossRef]
5. Blowman, K.; Magalhães, M.; Lemos, M.F.L.; Cabral, C.; Pires, I.M. Anticancer Properties of Essential Oils and Other Natural Products. *Evid.-Based Complement. Altern. Med.* **2018**, *2018*, 3149362. [CrossRef]
6. Mohamed Abdoul-Latif, F.; Ainane, A.; Houmed Aboubaker, I.; Mohamed, J.; Ainane, T. Exploring the Potent Anticancer Activity of Essential Oils and Their Bioactive Compounds: Mechanisms and Prospects for Future Cancer Therapy. *Pharmaceuticals* **2023**, *16*, 1086. [CrossRef]
7. Sharma, P.R.; Mondhe, D.M.; Muthiah, S.; Pal, H.C.; Shahi, A.K.; Saxena, A.K.; Qazi, G.N. Anticancer activity of an essential oil from *Cymbopogon flexuosus*. *Chem. Biol. Interact.* **2009**, *179*, 160–168. [CrossRef] [PubMed]
8. Gaonkar, R.; Shiralgi, Y.; Lakkappa, D.B.; Hegde, G. Essential oil from *Cymbopogon flexuosus* as the potential inhibitor for HSP90. *Toxicol. Rep.* **2018**, *5*, 489–496. [CrossRef]
9. Patel, P.B.; Thakkar, V.R.; Patel, J.S. Cellular Effect of Curcumin and Citral Combination on Breast Cancer Cells: Induction of Apoptosis and Cell Cycle Arrest. *J. Breast Cancer* **2015**, *18*, 225. [CrossRef]
10. Bailly, C. Targets and pathways involved in the antitumor activity of citral and its stereo-isomers. *Eur. J. Pharmacol.* **2020**, *871*, 172945. [CrossRef]
11. Viktorová, J.; Stupák, M.; Řehořová, K.; Dobiasová, S.; Hoang, L.; Hajšlová, J.; Van Thanh, T.; Van Tri, L.; Van Tuan, N.; Ruml, T. Lemon Grass Essential Oil does not Modulate Cancer Cells Multidrug Resistance by Citral—Its Dominant and Strongly Antimicrobial Compound. *Foods* **2020**, *9*, 585. [CrossRef]
12. Coricovac, D.; Dehelean, C.A.; Pinzaru, I.; Mioc, A.; Aburel, O.-M.; Macasoi, I.; Draghici, G.A.; Petean, C.; Soica, C.; Boruga, M.; et al. Assessment of Betulinic Acid Cytotoxicity and Mitochondrial Metabolism Impairment in a Human Melanoma Cell Line. *Int. J. Mol. Sci.* **2021**, *22*, 4870. [CrossRef] [PubMed]
13. Sheikh, B.Y.; Sarker, M.M.R.; Kamarudin, M.N.A.; Mohan, G. Antiproliferative and apoptosis inducing effects of citral via p53 and ROS-induced mitochondrial-mediated apoptosis in human colorectal HCT116 and HT29 cell lines. *Biomed. Pharmacother.* **2017**, *96*, 834–846. [CrossRef] [PubMed]
14. Sharma, M.; Grewal, K.; Jandrotia, R.; Batish, D.R.; Singh, H.P.; Kohli, R.K. Essential oils as anticancer agents: Potential role in malignancies, drug delivery mechanisms, and immune system enhancement. *Biomed. Pharmacother.* **2022**, *146*, 112514. [CrossRef]
15. Sanches, L.J.; Marinello, P.C.; Panis, C.; Fagundes, T.R.; Morgado-Díaz, J.A.; De-Freitas-Junior, J.C.M.; Cecchini, R.; Cecchini, A.L.; Luiz, R.C. Cytotoxicity of citral against melanoma cells: The involvement of oxidative stress generation and cell growth protein reduction. *Tumor Biol.* **2017**, *39*, 101042831769591. [CrossRef]
16. Younis, N.K.; Roumieh, R.; Bassil, E.P.; Ghoubaira, J.A.; Kobeissy, F.; Eid, A.H. Nanoparticles: Attractive tools to treat colorectal cancer. *Semin. Cancer Biol.* **2022**, *86*, 1–13. [CrossRef] [PubMed]
17. Lupea-Chilom, D.-S.; Solovan, C.S.; Farcas, S.S.; Gogulescu, A.; Andreescu, N.I. Latent Tuberculosis in Psoriasis Patients on Biologic Therapies: Real-World Data from a Care Center in Romania. *Medicina* **2023**, *59*, 1015. [CrossRef]
18. Burguin, A.; Diorio, C.; Durocher, F. Breast Cancer Treatments: Updates and New Challenges. *J. Pers. Med.* **2021**, *11*, 808. [CrossRef]

19. Talib, W.H.; Alsayed, A.R.; Barakat, M.; Abu-Taha, M.I.; Mahmod, A.I. Targeting Drug Chemo-Resistance in Cancer Using Natural Products. *Biomedicines* **2021**, *9*, 1353. [CrossRef]
20. Cruz-Martins, N. Advances in Plants-Derived Bioactives for Cancer Treatment. *Cells* **2023**, *12*, 1112. [CrossRef]
21. Mukarram, M.; Choudhary, S.; Khan, M.A.; Poltronieri, P.; Khan, M.M.A.; Ali, J.; Kurjak, D.; Shahid, M. Lemongrass Essential Oil Components with Antimicrobial and Anticancer Activities. *Antioxidants* **2021**, *11*, 20. [CrossRef]
22. Rahimi, G.; Yousefnia, S.; Angnes, L.; Negahdary, M. Design a PEGylated nanocarrier containing lemongrass essential oil (LEO), a drug delivery system: Application as a cytotoxic agent against breast cancer cells. *J. Drug Deliv. Sci. Technol.* **2023**, *80*, 104183. [CrossRef]
23. Basiouny, E.; Abdakaway, A.; Asker, M.; Abozid, M. A comparative study between three essential oils in terms of their chemical composition and antioxidant activity. *Menoufia J. Agric. Biotechnol.* **2023**, *8*, 81–91. [CrossRef]
24. Al Weshahi, H.; Akhtar, M.S.; Al Tobi, S.S.; Hossain, A.; Khan, S.A.; Akhtar, A.B.; Said, S.A. Evaluation of acute plant toxicity, antioxidant activity, molecular docking and bioactive compounds of lemongrass oil isolated from Omani cultivar. *Toxicol. Reports* **2025**, *14*, 101888. [CrossRef]
25. Ashaq, B.; Rasool, K.; Habib, S.; Bashir, I.; Nisar, N.; Mustafa, S.; Ayaz, Q.; Nayik, G.A.; Uddin, J.; Ramniwas, S.; et al. Insights into chemistry, extraction and industrial application of lemon grass essential oil—A review of recent advances. *Food Chem. X* **2024**, *22*, 101521. [CrossRef] [PubMed]
26. Wojtunik-Kulesza, K.A.; Cieśla, Ł.M.; Waksmundzka-Hajnos, M. Approach to Determination a Structure—Antioxidant Activity Relationship of Selected Common Terpenoids Evaluated by $ABTS^{\bullet+}$ Radical Cation Assay. *Nat. Prod. Commun.* **2018**, *13*, 295–299. [CrossRef]
27. Nirmala, M.J.; Durai, L.; Gopakumar, V.; Nagarajan, R. Anticancer and antibacterial effects of a clove bud essential oil-based nanoscale emulsion system. *Int. J. Nanomed.* **2019**, *14*, 6439–6450. [CrossRef] [PubMed]
28. Roozitalab, G.; Yousefpoor, Y.; Abdollahi, A.; Safari, M.; Rasti, F.; Osanloo, M. Antioxidative, anticancer, and antibacterial activities of a nanoemulsion-based gel containing *Myrtus communis* L. essential oil. *Chem. Pap.* **2022**, *76*, 4261–4271. [CrossRef]
29. Amina, M.; Al Musayeib, N.M.; Al-Hamoud, G.A.; Al-Dbass, A.; El-Ansary, A.; Ali, M.A. Prospective of biosynthesized *L. satiVum* oil/PEG/Ag-MgO bionanocomposite film for its antibacterial and anticancer potential. *Saudi J. Biol. Sci.* **2021**, *28*, 5971–5985. [CrossRef]
30. Cao, X.; Zhu, Q.; Wang, Q.-L.; Adu-Frimpong, M.; Wei, C.-M.; Weng, W.; Bao, R.; Wang, Y.-P.; Yu, J.-N.; Xu, X.M. Improvement of Oral Bioavailability and Anti-Tumor Effect of Zingerone Self-Microemulsion Drug Delivery System. *J. Pharm. Sci.* **2021**, *110*, 2718–2727. [CrossRef]
31. de Oliveira, M.C.; Bruschi, M.L. Self-Emulsifying Systems for Delivery of Bioactive Compounds from Natural Origin. *AAPS PharmSciTech* **2022**, *23*, 134. [CrossRef]
32. Milan, A.; Mioc, M.; Mioc, A.; Gogulescu, A.; Mardale, G.; Avram, Ș.; Maksimović, T.; Mara, B.; Șoica, C. Cytotoxic Potential of Betulinic Acid Fatty Esters and Their Liposomal Formulations: Targeting Breast, Colon, and Lung Cancer Cell Lines. *Molecules* **2024**, *29*, 3399. [CrossRef] [PubMed]
33. Anand, U.; Dey, A.; Chandel, A.K.S.; Sanyal, R.; Mishra, A.; Pandey, D.K.; De Falco, V.; Upadhyay, A.; Kandimalla, R.; Chaudhary, A.; et al. Cancer chemotherapy and beyond: Current status, drug candidates, associated risks and progress in targeted therapeutics. *Genes Dis.* **2023**, *10*, 1367–1401. [CrossRef] [PubMed]
34. Al-Ghanayem, A.A. Antifungal Activity of *Cymbopogon flexuosus* Essential Oil and its Effect on Biofilm Formed by *Candida parapsilosis* and *Candida tropicalis* on Polystyrene and Polyvinyl Plastic Surfaces. *Indian J. Pharm. Educ. Res.* **2023**, *57*, 113–119. [CrossRef]
35. Adukwu, E.C.; Bowles, M.; Edwards-Jones, V.; Bone, H. Antimicrobial activity, cytotoxicity and chemical analysis of lemongrass essential oil (*Cymbopogon flexuosus*) and pure citral. *Appl. Microbiol. Biotechnol.* **2016**, *100*, 9619–9627. [CrossRef]
36. Kumar, A.; Malik, F.; Bhushan, S.; Sethi, V.K.; Shahi, A.K.; Kaur, J.; Taneja, S.C.; Qazi, G.N.; Singh, J. An essential oil and its major constituent isointermedeol induce apoptosis by increased expression of mitochondrial cytochrome c and apical death receptors in human leukaemia HL-60 cells. *Chem. Biol. Interact.* **2008**, *171*, 332–347. [CrossRef]
37. Wang, Y.-F.; Zheng, Y.; Cha, Y.-Y.; Feng, Y.; Dai, S.-X.; Zhao, S.; Chen, H.; Xu, M. Essential oil of lemon myrtle (*Backhousia citriodora*) induces S-phase cell cycle arrest and apoptosis in HepG2 cells. *J. Ethnopharmacol.* **2023**, *312*, 116493. [CrossRef]
38. Luang-In, V.; Saengha, W.; Karirat, T.; Senakun, C.; Siriamornpun, S. Phytochemical Profile of Cymbopogon citratus (DC.) Stapf Lemongrass Essential Oil from Northeastern Thailand and Its Antioxidant and Antimicrobial Attributes and Cytotoxic Effects on HT-29 Human Colorectal Adenocarcinoma Cells. *Foods* **2024**, *13*, 2928. [CrossRef]
39. Srivastava, G.; Mukherjee, E.; Mittal, R.; Ganjewala, D. Geraniol and citral: Recent developments in their anticancer credentials opening new vistas in complementary cancer therapy. *Z. Für Naturforsch. C* **2024**, *79*, 163–177. [CrossRef]
40. Silva, B.I.M.; Nascimento, E.A.; Silva, C.J.; Silva, T.G.; Aguiar, J.S. Anticancer activity of monoterpenes: A systematic review. *Mol. Biol. Rep.* **2021**, *48*, 5775–5785. [CrossRef]

41. Lee, J.-E.; Seo, S.-M.; Huh, M.-J.; Lee, S.-C.; Park, I.-K. Reactive oxygen species mediated-antifungal activity of cinnamon bark (*Cinnamomum verum*) and lemongrass (*Cymbopogon citratus*) essential oils and their constituents against two phytopathogenic fungi. *Pestic. Biochem. Physiol.* **2020**, *168*, 104644. [CrossRef]
42. Najar, B.; Shortrede, J.E.; Pistelli, L.; Buhagiar, J. Chemical Composition and In Vitro Cytotoxic Screening of Sixteen Commercial Essential Oils on Five Cancer Cell Lines. *Chem. Biodivers.* **2020**, *17*, e1900478. [CrossRef] [PubMed]
43. Alotaibi, M.A. Lemongrass Extract and Anticancer Impact: mRNA Levels of Expressing Apoptosis and Mitochondrial Fission in MCF7 Cells. *Pak. J. Zool.* **2024**, *56*, 1–7. [CrossRef]
44. Alwaili, M.A. Protective effects of lemongrass (*Cymbopogon citratus* STAPF) extract mediated mitochondrial fission and glucose uptake inhibition in SW1417. *Food Sci. Technol.* **2023**, *43*, e94522. [CrossRef]
45. Balaban, R.S.; Nemoto, S.; Finkel, T. Mitochondria, Oxidants, and Aging. *Cell* **2005**, *120*, 483–495. [CrossRef]
46. Sabharwal, S.S.; Schumacker, P.T. Mitochondrial ROS in cancer: Initiators, amplifiers or an Achilles' heel? *Nat. Rev. Cancer* **2014**, *14*, 709–721. [CrossRef]
47. Pelicano, H.; Lu, W.; Zhou, Y.; Zhang, W.; Chen, Z.; Hu, Y.; Huang, P. Mitochondrial Dysfunction and Reactive Oxygen Species Imbalance Promote Breast Cancer Cell Motility through a CXCL14-Mediated Mechanism. *Cancer Res.* **2009**, *69*, 2375–2383. [CrossRef]
48. Nakazato, T.; Ito, K.; Ikeda, Y.; Kizaki, M. Green Tea Component, Catechin, Induces Apoptosis of Human Malignant B Cells via Production of Reactive Oxygen Species. *Clin. Cancer Res.* **2005**, *11*, 6040–6049. [CrossRef]
49. Du, J.; Daniels, D.H.; Asbury, C.; Venkataraman, S.; Liu, J.; Spitz, D.R.; Oberley, L.W.; Cullen, J.J. Mitochondrial Production of Reactive Oxygen Species Mediate Dicumarol-induced Cytotoxicity in Cancer Cells. *J. Biol. Chem.* **2006**, *281*, 37416–37426. [CrossRef] [PubMed]
50. Zheng, S.; Jing, G.; Wang, X.; Ouyang, Q.; Jia, L.; Tao, N. Citral exerts its antifungal activity against *Penicillium digitatum* by affecting the mitochondrial morphology and function. *Food Chem.* **2015**, *178*, 76–81. [CrossRef]
51. OuYang, Q.; Tao, N.; Zhang, M. A Damaged Oxidative Phosphorylation Mechanism Is Involved in the Antifungal Activity of Citral against *Penicillium digitatum*. *Front. Microbiol.* **2018**, *9*, 239. [CrossRef]
52. Munteanu, A.; Gogulescu, A.; Șoica, C.; Mioc, A.; Mioc, M.; Milan, A.; Lukinich-Gruia, A.T.; Pricop, M.-A.; Jianu, C.; Banciu, C.; et al. In Vitro and In Silico Evaluation of Syzygium aromaticum Essential Oil: Effects on Mitochondrial Function and Cytotoxic Potential Against Cancer Cells. *Plants* **2024**, *13*, 3443. [CrossRef] [PubMed]
53. Bezerra, D.; Militão, G.; De Morais, M.; De Sousa, D. The Dual Antioxidant/Prooxidant Effect of Eugenol and Its Action in Cancer Development and Treatment. *Nutrients* **2017**, *9*, 1367. [CrossRef] [PubMed]
54. Park, K.-R.; Nam, D.; Yun, H.-M.; Lee, S.-G.; Jang, H.-J.; Sethi, G.; Cho, S.K.; Ahn, K.S. β-Caryophyllene oxide inhibits growth and induces apoptosis through the suppression of PI3K/AKT/mTOR/S6K1 pathways and ROS-mediated MAPKs activation. *Cancer Lett.* **2011**, *312*, 178–188. [CrossRef] [PubMed]
55. Ouvry, G.; Clary, L.; Tomas, L.; Aurelly, M.; Bonnary, L.; Borde, E.; Bouix-Peter, C.; Chantalat, L.; Defoin-Platel, C.; Deret, S.; et al. Impact of Minor Structural Modifications on Properties of a Series of mTOR Inhibitors. *ACS Med. Chem. Lett.* **2019**, *10*, 1561–1567. [CrossRef]
56. Craveiro, A.A. Oleos essenciais de plantas do Nordeste. In *Coleção Ciência*; Edições UFC: Fortaleza, Brazil, 1981; p. 209.
57. Craveiro, A.A.; Matos, F.J.A.; de Alencar, J.W. A simple and inexpensive steam generator for essential oils extraction. *J. Chem. Educ.* **1976**, *53*, 652. [CrossRef]
58. Jianu, C.; Rusu, L.-C.; Muntean, I.; Cocan, I.; Lukinich-Gruia, A.T.; Goleț, I.; Horhat, D.; Mioc, M.; Mioc, A.; Șoica, C.; et al. In Vitro and In Silico Evaluation of the Antimicrobial and Antioxidant Potential of *Thymus pulegioides* Essential Oil. *Antioxidants* **2022**, *11*, 2472. [CrossRef]
59. Rădulescu, M.; Jianu, C.; Lukinich-Gruia, A.T.; Mioc, M.; Mioc, A.; Șoica, C.; Stana, L.G. Chemical Composition, In Vitro and In Silico Antioxidant Potential of *Melissa officinalis* subsp. *officinalis Essential Oil*. *Antioxidants* **2021**, *10*, 1081. [CrossRef]
60. Pricop, M.-A.; Lukinich-Gruia, A.T.; Cristea, I.-M.; Păunescu, V.; Tatu, C.A. Aristolochia clematitis L. Ethanolic Extracts: In Vitro Evaluation of Antioxidant Activity and Cytotoxicity on Caco-2 Cell Line. *Plants* **2024**, *13*, 2987. [CrossRef]
61. Abcam, C. (UK): DCFDA/H2DCFDA—Cellular ROS Assay Kit (ab113851). Available online: https://doc.abcam.com/datasheets/active/ab113851/en-us/dcfda-h2dcfda-cellular-ros-assay-kit-ab113851.pdf (accessed on 3 December 2024).
62. Abcam, C. (UK): JC-1 Mitochondrial Membrane Potential Assay Kit (ab113850). Available online: https://www.abcam.com/en-ro/products/assay-kits/jc-1-mitochondrial-membrane-potential-assay-kit-ab113850 (accessed on 3 November 2024).
63. Petrus, A.T.; Lighezan, D.L.; Danila, M.D.; Duicu, O.M.; Sturza, A.; Muntean, D.M.; Ionita, I. Assessment of Platelet Respiration as Emerging Biomarker of Disease. *Physiol. Res.* **2019**, *68*, 347–363. [CrossRef]
64. Berman, H.M. The Protein Data Bank. *Nucleic Acids Res.* **2000**, *28*, 235–242. [CrossRef]
65. Kim, S.; Chen, J.; Cheng, T.; Gindulyte, A.; He, J.; He, S.; Li, Q.; Shoemaker, B.A.; Thiessen, P.A.; Yu, B.; et al. PubChem 2023 update. *Nucleic Acids Res.* **2023**, *51*, D1373–D1380. [CrossRef] [PubMed]

66. Dallakyan, S.; Olson, A.J. Small-Molecule Library Screening by Docking with PyRx. In *Chemical Biology: Methods and Protocols*; Springer: New York, NY, USA, 2015; pp. 243–250.
67. Trott, O.; Olson, A.J. AutoDock Vina: Improving the speed and accuracy of docking with a new scoring function, efficient optimization, and multithreading. *J. Comput. Chem.* **2010**, *31*, 455–461. [CrossRef] [PubMed]
68. McTigue, M.; Murray, B.W.; Chen, J.H.; Deng, Y.-L.; Solowiej, J.; Kania, R.S. Molecular conformations, interactions, and properties associated with drug efficiency and clinical performance among VEGFR TK inhibitors. *Proc. Natl. Acad. Sci. USA* **2012**, *109*, 18281–18289. [CrossRef]
69. Wood, E.R.; Truesdale, A.T.; McDonald, O.B.; Yuan, D.; Hassell, A.; Dickerson, S.H.; Ellis, B.; Pennisi, C.; Horne, E.; Lackey, K.; et al. A Unique Structure for Epidermal Growth Factor Receptor Bound to GW572016 (Lapatinib). *Cancer Res.* **2004**, *64*, 6652–6659. [CrossRef] [PubMed]
70. Tecle, H.; Shao, J.; Li, Y.; Kothe, M.; Kazmirski, S.; Penzotti, J.; Ding, Y.-H.; Ohren, J.; Moshinsky, D.; Coli, R.; et al. Beyond the MEK-pocket: Can current MEK kinase inhibitors be utilized to synthesize novel type III NCKIs? Does the MEK-pocket exist in kinases other than MEK? *Bioorg. Med. Chem. Lett.* **2009**, *19*, 226–229. [CrossRef]
71. Islam, I.; Bryant, J.; Chou, Y.-L.; Kochanny, M.J.; Lee, W.; Phillips, G.B.; Yu, H.; Adler, M.; Whitlow, M.; Ho, E.; et al. Indolinone based phosphoinositide-dependent kinase-1 (PDK1) inhibitors. Part 1: Design, synthesis and biological activity. *Bioorg. Med. Chem. Lett.* **2007**, *17*, 3814–3818. [CrossRef]
72. Addie, M.; Ballard, P.; Buttar, D.; Crafter, C.; Currie, G.; Davies, B.R.; Debreczeni, J.; Dry, H.; Dudley, P.; Greenwood, R.; et al. Discovery of 4-Amino-N-[(1S)-1-(4-chlorophenyl)-3-hydroxypropyl]-1-(7H-pyrrolo[2,3-d]pyrimidin-4-yl)piperidine-4-carboxamide (AZD5363), an Orally Bioavailable, Potent Inhibitor of Akt Kinases. *J. Med. Chem.* **2013**, *56*, 2059–2073. [CrossRef]
73. Le, P.T.; Cheng, H.; Ninkovic, S.; Plewe, M.; Huang, X.; Wang, H.; Bagrodia, S.; Sun, S.; Knighton, D.R.; LaFleur Rogers, C.M.; et al. Design and synthesis of a novel pyrrolidinyl pyrido pyrimidinone derivative as a potent inhibitor of PI3Kα and mTOR. *Bioorg. Med. Chem. Lett.* **2012**, *22*, 5098–5103. [CrossRef]
74. Yang, H.; Rudge, D.G.; Koos, J.D.; Vaidialingam, B.; Yang, H.J.; Pavletich, N.P. mTOR kinase structure, mechanism and regulation. *Nature* **2013**, *497*, 217–223. [CrossRef]
75. Lee, E.F.; Czabotar, P.E.; Smith, B.J.; Deshayes, K.; Zobel, K.; Colman, P.M.; Fairlie, W.D. Crystal structure of ABT-737 complexed with Bcl-xL: Implications for selectivity of antagonists of the Bcl-2 family. *Cell Death Differ.* **2007**, *14*, 1711–1713. [CrossRef]
76. Souers, A.J.; Leverson, J.D.; Boghaert, E.R.; Ackler, S.L.; Catron, N.D.; Chen, J.; Dayton, B.D.; Ding, H.; Enschede, S.H.; Fairbrother, W.J.; et al. ABT-199, a potent and selective BCL-2 inhibitor, achieves antitumor activity while sparing platelets. *Nat. Med.* **2013**, *19*, 202–208. [CrossRef] [PubMed]

plants

MDPI

Article

Herbicidal Formulations with Plant-Based Compounds to Control *Amaranthus hybridus*, *Lolium multiflorum*, and *Brassica rapa* Weeds

Juan J. Romero [1,2], Juliana Soler-Arango [1,2], Marcos E. Coustet [1], Daniela B. Moracci [1], Sebastián Reinoso [1], Marcos E. Yanniccari [1,2,3,*], Aline Schneider-Teixeira [1,2] and Jimena M. Herrera [2,4]

1 YPF Tecnología (Y-TEC), Av. del Petróleo S/N entre 129 y 143, Berisso 1923, Argentina; juanjose.romero@ypftecnologia.com (J.J.R.); juliana.solerarango@ypftecnologia.com (J.S.-A.); marcos.coustet@ypftecnologia.com (M.E.C.); daniela.b.moracci@set.ypf.com (D.B.M.); sebastian.reinoso@ypftecnologia.com (S.R.); aline.s.teixeira@ypftecnologia.com (A.S.-T.)
2 Consejo Nacional de Investigaciones Científicas y Técnicas (CONICET), Godoy Cruz 2290, Ciudad Autónoma de Buenos Aires 5000, Argentina; jimenita_herrera@yahoo.com.ar
3 Chacra Experimental Integrada Barrow (MDA—INTA), Ruta 3. Km 487.5, Tres Arroyos 7500, Argentina
4 Instituto Multidisciplinario de Biología Vegetal (CONICET—UNC), Av. Vélez Sarsfield 1611, Córdoba 5016, Argentina
* Correspondence: marcosyanniccari@conicet.gov.ar; Tel.: +54-2983-43-1083 (ext. 125)

Abstract: Numerous studies have shown the potential effect of bioactive agents against weeds. In this study, we developed two binary formulations with nonanoic acid, citral, or thymoquinone as herbicides and evaluated their physicochemical properties. The presence of the bioactive compounds in the formulations was confirmed through FTIR spectroscopy. A dynamic light scattering study was conducted to characterize the emulsified formulations and the size and distribution of the aggregates. In addition, thermogravimetric analysis was performed to ensure the thermal stability of the formulations. The herbicidal activity against *Amaranthus hybridus*, *Lolium multiflorum*, and *Brassica rapa* weeds was evaluated, and each species showed different levels of sensitivity with half maximal inhibitory concentration doses from 0.07 to 5 mM. The binary formulations negatively affected the photosynthetic system reducing Fv/Fm values at 5 days after treatment. Lastly, the phytotoxic effect of the formulations was tested on wheat germination, and they did not inhibit plant germination and seedling growth at $\leq$5 mM after 14 days of application. The development of new formulations with natural compounds as bioactive ingredients would allow control of a wide spectrum of weeds through a multitarget-site effect.

Keywords: binary formulations; bioactive compounds; herbicide; natural compounds; weeds

1. Introduction

The world population growth projections indicate that higher levels of food production will be required [1]; therefore, agricultural best management practices represent a critical aspect of production systems. *Amaranthus hybridus*, *Lolium multiflorum*, and *Brassica rapa* are common weed species in Argentina and are present in 40% of crop fields [2]. *A. hybridus* is an annual Amaranthaceae weed found in tropical and subtropical regions and produces a large quantity of seeds easily dispersed, *B. rapa* is an annual Brassicaceae weed found in all continents, *L. multiflorum* is an annual weed found in large part of the Pampas region, and the three species have evolved herbicide resistance [2–7]. Anyway, the management of these weeds continues to be based on chemical control [2]. However, the increase in herbicide usage has been associated with environmental and human health risks [8]. Currently, some herbicides such as atrazine have been listed as restricted in some countries [9]. An alternative to weed control is the use of plant-based compounds. These are considered of minimum risk due to their low toxicity, easy degradability in the environment, and non-restricted use [9].

Numerous studies have shown the potential effect of botanical agents, such as fatty acids, aldehydes, ketones, terpenes, and others, against weeds [10,11]. The phytotoxic effect of essential oils (EOs) and their pure components have been studied. For example, the herbicidal activity of pure terpenes inhibits the germination and seedling growth in *Sinapis arvensis*, *Amaranthus retroflexus*, *Centaurea solstitialis*, *Raphanus raphanistrum*, *Rumex nepalensis*, and *Sonchus oleraceus*. Kordali et al. [12] and Azirak and Karaman [13] reported that the herbicidal effect of pure terpenes was higher than that of the commercial herbicide 2,4-D isooctyl ester. The effect of these terpenes has been attributed to their interference with plant cell processes, including mitosis inhibition and decreased cellular respiration and chlorophyll content [11].

Nonanoic acid is a saturated fatty acid derived from *Pelargonium* spp. EO, and it is used as a post-emergent weed control agent [14]. Citral (geranial and neral) is an aldehyde terpene found in the EOs of several aromatic plants, such as *Cymbopogon citratus*, *Citrus* sp., and *Lippia* sp. [15,16], and is a weed germination inhibitor [17,18]. Thymoquinone is one of the main active compounds of the EO from black cumin (*Nigella sativa*) and exhibits a wide range of activities such as pharmacological and biopesticide activities [19,20]. These active compounds are also used in the food industry as natural antimicrobials and additives for food preservation [21].

The objective of the current work was to develop binary formulations based on bioactive compounds with herbicidal or biopesticide activities. An emulsifier was used due to the volatility of the compounds [18]. In addition, physicochemical properties and herbicidal activity against weeds were evaluated. Thus, the effect of the binary formulations on PSII was tested. Lastly, the phytotoxic effect of the formulations was evaluated on wheat germination and seedling growth. Therefore, the development of new formulations with different bioactive ingredients would allow control of a wide spectrum of weeds through a multitarget-site effect and new modes of action while causing a low impact on the environment.

2. Materials and Methods

2.1. Plant Materials

Amaranthus hybridus and *Brassica rapa* seeds were collected from weed populations naturalized in crop fields from Tres Arroyos, Argentina (Figure 1A,B). Both species are the most relevant herbicide-resistant weeds in Argentina; they have evolved resistance to multiple herbicides (glyphosate and acetolactate synthase inhibiting herbicides) [2] and frequently reach high density in crop fields (Figure 1A,B). *Lolium multiflorum* was supplied by Barenbrung, Pergamino, Buenos Aires, Argentina. Seeds of *Lactuca sativa* L. var. *crispa* and *Solanum lycopersicum* L. var. *platense* were used as model plants and obtained from Facultad de Ciencias Agrarias y Forestales, Universidad Nacional de La Plata, Argentina. Wheat seeds of var. Tero were provided by Illinois, Argentina.

Figure 1. Multiple herbicide-resistant weed populations harvested to obtain seeds employed in the experiments: (**A**) *Amaranthus hybridus* in soybean crop and (**B**) *Brassica rapa* in oat crop.

2.2. Chemical Compounds

The bioactive compounds nonanoic acid (96%, catalog number 807167), citral (96%, catalog number 8024890250), and thymoquinone (98%, catalog number 274666) were analytical grade and purchased from Sigma-Aldrich Chemical Co. (Steinheim, Germany) (Table 1). The soy lecithin emulsifier (35% p/p) and atrazine (positive control, 90%) were provided by YPF S.A., Capital Federal, Buenos Aires, Argentina

Table 1. Bioactive compounds and physicochemical properties.

Chemical Structures	Boiling Point (°C)	Log P	Solubility in Water at 25 °C (g/L)
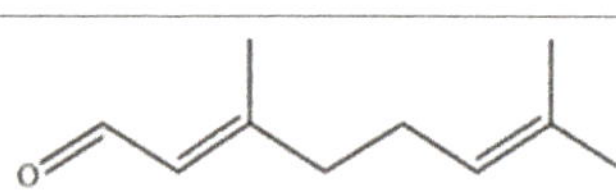 citral	225	3.17	0.08

Table 1. *Cont.*

Chemical Structures	Boiling Point (°C)	Log P	Solubility in Water at 25 °C (g/L)
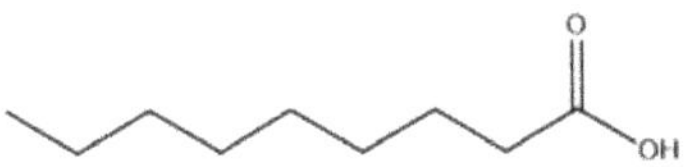 nonanoic acid	254	3.43	0.21
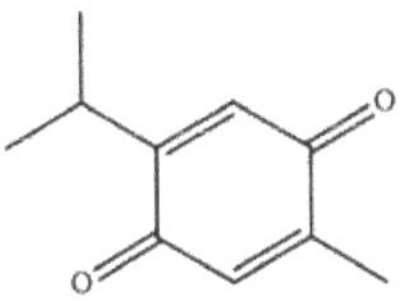 thymoquinone	230–232	2.33	0.87

2.3. Formulations

Two binary formulations (F1 and F2) were developed as bioherbicides. F1 contains nonanoic acid and citral (1:1), while F2 contains nonanoic acid and thymoquinone (1:1). Due to the physiochemical properties of the bioactive agents [22] (Table 1), an emulsifier was added.

2.4. Chemical Characterization

2.4.1. Fourier Transform Infrared Spectrometry (ATR-FTIR)

A molecular interaction study was conducted using ATR-FTIR. Spectra were recorded with a Nicolet™ iS™10 (Thermo Scientific, Madison, WI, USA). The spectrum of each sample was the average of three successive scans in the 4000–400 cm^{-1} wavenumber range and recorded as the absorbance (calculated as the logarithm of the reflectance reciprocal).

To analyze F1 and F2 binary formulations, a drop was placed on the diamond ATR crystal using a top plate and pressure arm accessories (Smart iTX accessory, Madison, WI, USA). A spectral analysis was performed with the software Omnic version 9 (Thermo Scientific).

2.4.2. Dynamic Light Scattering (DLS) Measurements

The emulsified formulations, size, and size distribution of aggregates were determined through DLS as a function of temperature using a Zetasizer Nano-ZS90 (Malvern Panalytical, Malvern, UK). Samples were properly dispersed in distilled water at room temperature before their analysis to avoid multiple scattering effects. The experiments were performed with emulsified formulations dispersed in distilled water at 25, 35, and 45 °C, typical summer temperatures in Argentina.

2.4.3. Thermogravimetric Analysis (TGA)

To ensure thermal stability, TGA was performed using a TGA Discovery 5500 analyzer (TA Instruments, New Castle, DE, USA), under nitrogen flux (20 mL min^{-1}), in platinum pans heated up to 150 °C with a 10 °C min^{-1} rate and isothermal steps at 25, 35, and 45 °C, typical summer temperatures in Argentina.

2.5. Herbicide Activity

2.5.1. Weed Germination Inhibition Test

The herbicidal activity of F1 and F2 was tested against weed and model plant seeds following the methodology described by Sosa et al. [23] with some modifications. Briefly,

4 mL of aqueous emulsions containing F1 or F2 at 0.07 to 5 mM were tested, and an emulsifier was added to the formulation at 0.2% v/v [24]. The extracts were placed onto 9 cm diameter paper disks in Petri dishes. Then, 10 seeds of each species were placed onto paper disks. Subsequently, the dishes were closed under the following experimental conditions: room temperature, 26.0 ± 1.8 °C; relative humidity, $50 \pm 9.7\%$; and photoperiod, 12:12. At 7 days after sowing, the seeds were considered germinated if their roots were longer than 1 mm [25]. Emulsions without the addition of the bioactive formulation were used as a negative control, whereas atrazine was used as a positive control because it targets a broad spectrum of weeds (monocotyledons and dicotyledons) [24,26]. The assays were performed in triplicates for each concentration. Half maximal inhibitory concentrations (IC50) were determined after 7 days of exposure and were calculated using POLO PLUS 2002–2007 LeOra Software [27].

2.5.2. Effect of Binary Formulations on Photosystem II (PSII)

The damage on PSII of the formulations was evaluated on seedlings of *A. hybridus*, because it is a main weed in a wide variety of crops worldwide [28], following the methodology described by Pooja et al. [29] with modifications. Briefly, 25 seedlings with 7 days post-emergence were placed onto paper filters into plastic trays (15.5 cm × 11 cm × 4 cm). Then, 10 mL of aqueous emulsions containing F1 or F2 at 0.07 to 5 mM and an emulsifier incorporated into the extracts at 0.2% v/v were added into boxes. Aqueous extracts without the addition of the bioactive formulation were used as a negative control, whereas atrazine (5 mM) was used as a positive control. In order to avoid possible chlorosis symptoms due to a lack of nutrients, 5 mL of nutritive solution (5.0 mM Ca (NO_3) 2, 5.0 mM KNO_3, 2.0 mM $MgSO_4$, 1.0 mM KH_2PO_4, 20.0 µM FeNa EDTA, 5.0 µM H_3BO_3, 0.9 µM $MnCl_2$, 0.8 µM $ZnCl_2$, 0.3 µM $CuSO_4$ y 0.01 µM Na_2MoO_4, pH 5.5–6.5) was added to each tray. The boxes were placed in a culture chamber at 22 ± 2 °C and 50 µmol photons m^{-2} s^{-1} with a photoperiod of 16 h. The chlorophyll fluorescence measures (Fv/Fm) were carried out with a MINI-PAM II ©Walz fluorimeter (Effeltrich, Germany) at 1, 2, 5, 7, and 12 days after application. The seedlings were darkened 30 min before measurements were taken.

For the experiments, three repetitions per treatment were used: Both experiments had four treatments. Each experiment was replicated three times. The treatments were assigned randomly to each box following a completely randomized design. The statistical analysis for Fv/Fm data was performed using InfoStat 2008 Software through Generalized Mixed Linear Models [30].

2.5.3. Effect of Binary Formulations on Wheat Germination and Seedling Growth

Effects of F1 and F2 on wheat germination and seedling growth were evaluated. The experiments were conducted according to Peschiutta et al. [31] with some modifications. F1 and F2 were added to Petri dishes at 2 and 5 mM and after 1, 7, or 14 days. Ten wheat seeds were placed in the Petri dishes. A negative control (H_2O) treatment was performed without the addition of any active compound, and the seeds were placed onto paper disks at the same time intervals. Then, the dishes were closed under the same experimental conditions mentioned above. The number of germinated seeds per dish was recorded 7 days after sowing. The seeds were considered germinated if their roots were longer than 1 mm. At that time, the leaf length of the wheat seedlings was measured. The assay was performed in triplicates. Data were analyzed to assess normality using the Shapiro–Wilk test, and homogeneity of the variances was determined using Levene's test before performing ANOVA. Tukey's tests were used to compare the means for germination and leaf length of seedlings between treatments through Infostat Software [30].

3. Results

3.1. Chemical Characterization

3.1.1. Fourier Transform Infrared Spectrometry (ATR-FTIR)

Figures 2 and 3 show the typical FTIR spectra of components corresponding to herbicidal formulations called F1 and F2, respectively. The characteristic band at 3343 cm^{-1} (grey area) that appears for all formulations is attributed to the O-H stretching vibration from the nonanoic acid and emulsifier. In addition, another band from the emulsifier is located at 3008 cm^{-1} (blue area) for F1 (Figure 2), related to asymmetric and symmetric C-H stretching vibrations, whereas the band for F2 has two origins, emulsifier and the thymoquinone, related to =C-H stretching vibrations (Figure 3). The spectrum shows the characteristic bands at 2923.2 cm^{-1} (orange area) associated with the C-H ester stretching vibration, the band at 1675.5 cm^{-1} (red area) is due to the stretching vibration of the ketone and aldehyde C=O group, and the band at 1117 cm^{-1} (green area) is attributed to the ether (C-O) groups. The last three bands mentioned are present in all the F1 and F2 components.

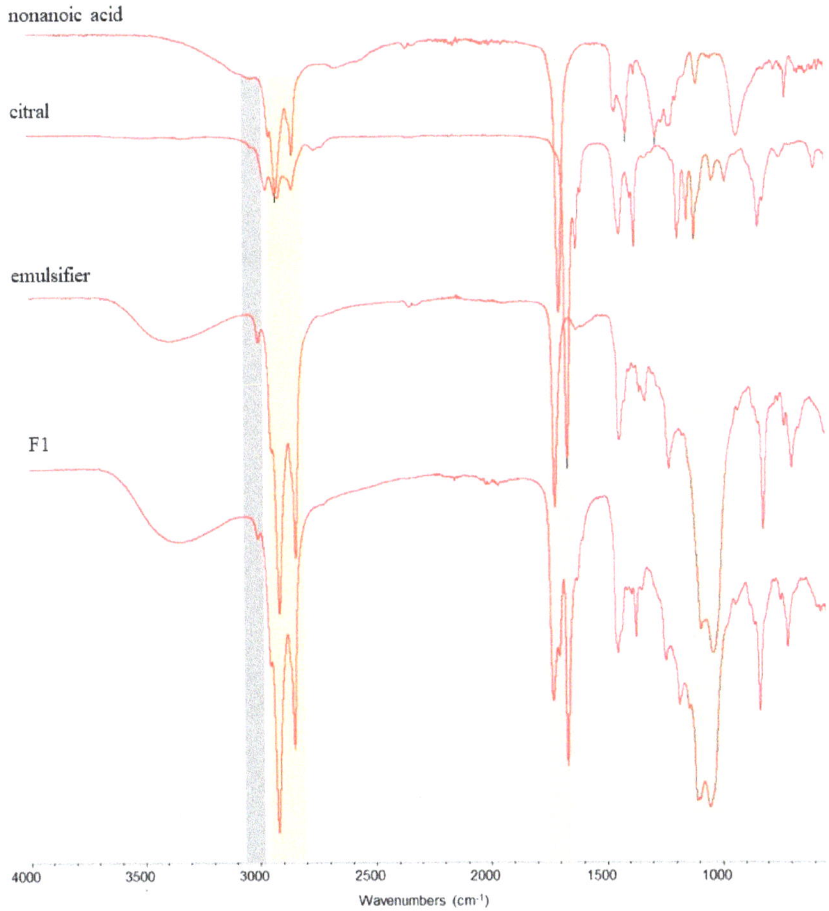

Figure 2. FTIR spectra of components (nonanoic acid, citral, and emulsifier) corresponding to formulation F1. The colored areas represent the bands of the oxygenated organic compound functional groups of interest: grey, OH; grey, =C-H; orange, C-H; red, C=O; and green, C-O.

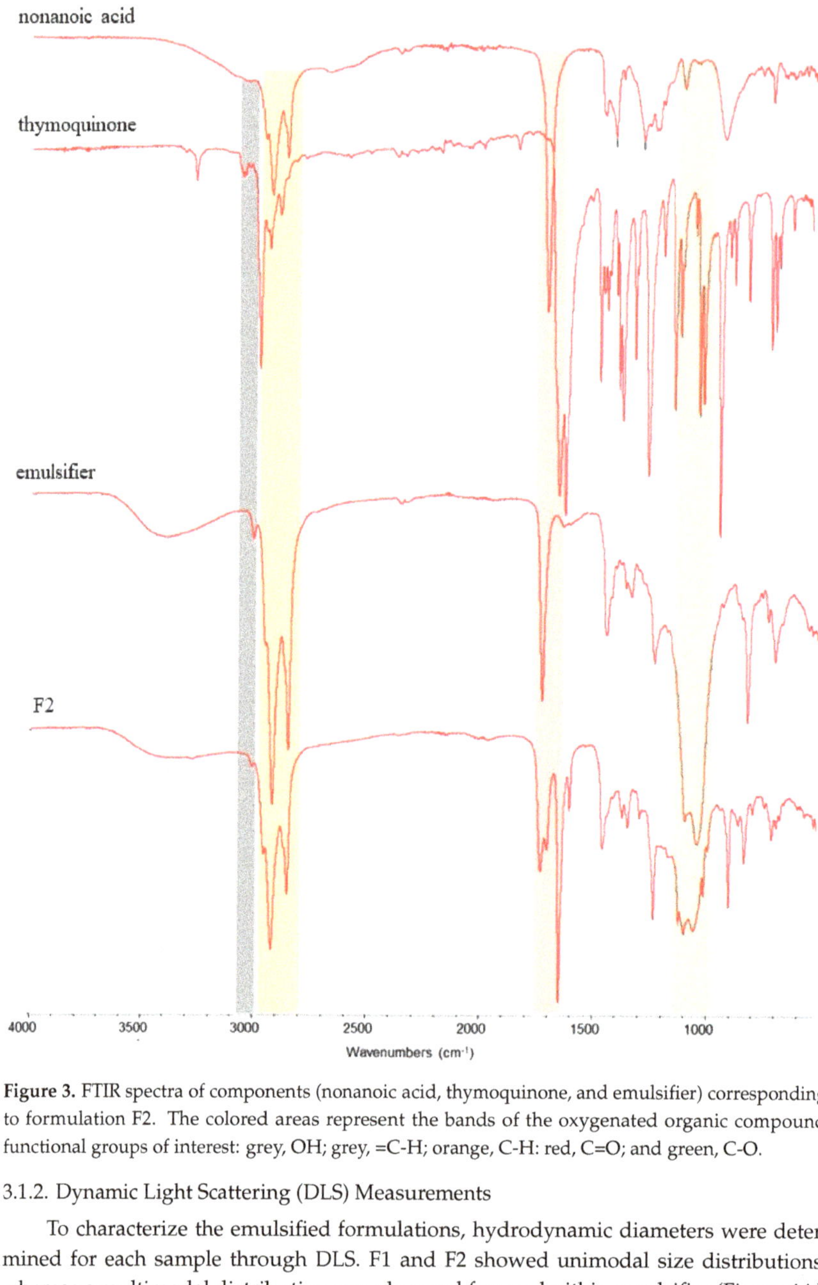

Figure 3. FTIR spectra of components (nonanoic acid, thymoquinone, and emulsifier) corresponding to formulation F2. The colored areas represent the bands of the oxygenated organic compound functional groups of interest: grey, OH; grey, =C-H; orange, C-H: red, C=O; and green, C-O.

3.1.2. Dynamic Light Scattering (DLS) Measurements

To characterize the emulsified formulations, hydrodynamic diameters were determined for each sample through DLS. F1 and F2 showed unimodal size distributions, whereas a multimodal distribution was observed for soy lecithin emulsifier (Figure 4A). Thus, for the temperature range involved in the experiments, the average sizes were from 800–1300 nm for the emulsifier, while F1 and F2 showed dispersion sizes of 250–550 and 170–260 nm, respectively.

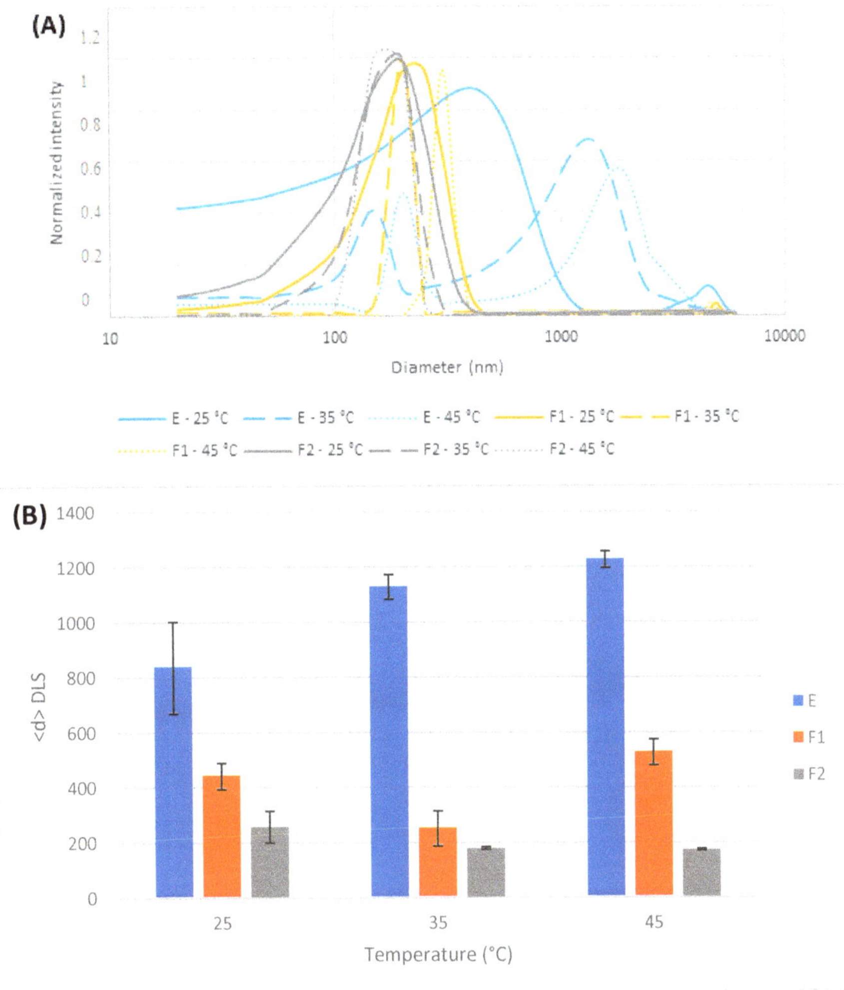

Figure 4. (**A**) DLS size distribution for emulsifier (E, blue), formulation 1 (F1, orange), and formulation 2 (F2, grey) at 25 (line), 35 (slashed), and 45 °C (dotted). (**B**) Temperature effect on averaged diameter (d) for emulsifier (E, blue), F1 (orange), and F2 (grey).

The presence of nonpolar analytes (active compounds) in the formulation turned the E dispersions into clear emulsions with unimodal distributions and lower polydispersity indexes. In contrast, the observed change in F1 and F2 size distributions with temperature was in line with the expected values for stable colloidal systems, with no flocculating aggregates under our experimental conditions (Figure 4B).

3.1.3. Thermogravimetric Analysis (TGA)

TGA was used to assess the thermal stability of the formulations. Thermograms of pure active compounds (thymoquinone, citral, and nonanoic acid), emulsifier, F1, and F2 were obtained for comparison. Weight losses of less than 10% were observed at temperatures under 100 °C for pure active compounds, except thymoquinone, which lost 17.5% of its weight. According to previously reported thermal assays (Figure 5A), this behavior can be the result of evaporation processes, considering the boiling point of these compounds and the constant N_2 flux (Table 1). In contrast, Figure 5A shows that the formulation composition did not affect significantly the recorded weight losses. In these cases, the presence of water was responsible for the considerable weight losses at temperatures under 100 °C, and the final weight loss corresponded to the remaining amount of the sample, consisting of active compounds and the emulsifier in the formulations. Lastly, the active compounds showed adequate thermal stability at all temperatures tested.

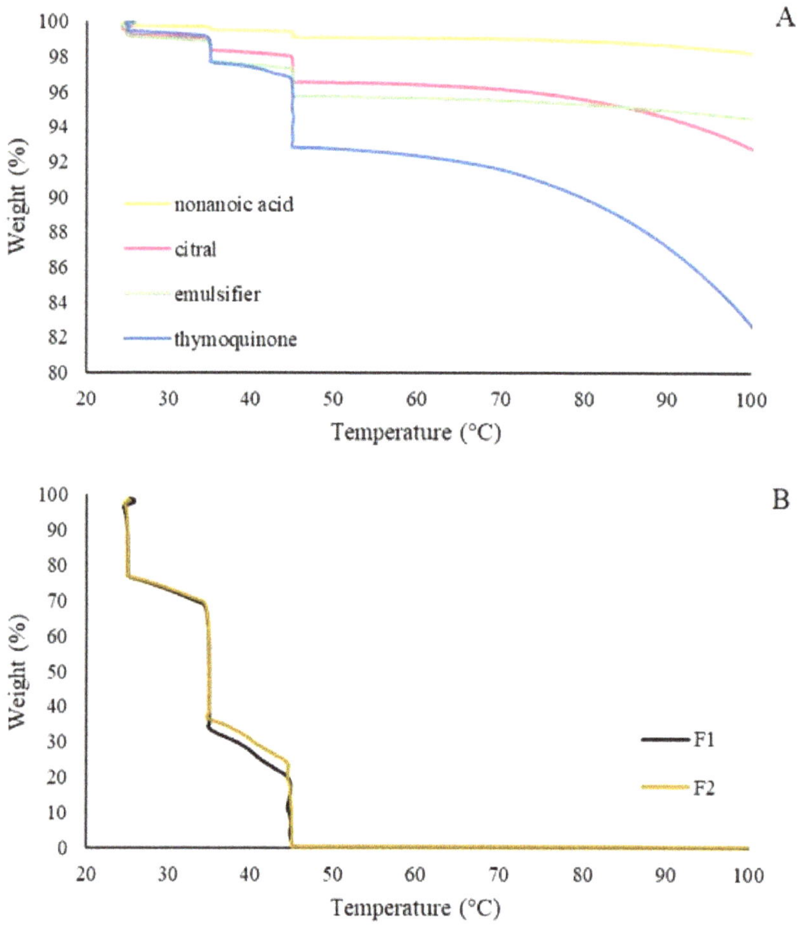

Figure 5. TGA thermograms: (**A**) pure active compounds (nonanoic acid, citral, thymoquinone, and emulsifier) and (**B**) formulations (F1 and F2).

3.2. Herbicide Activity

3.2.1. Weed Germination Inhibition Test

Germination inhibition of weeds induced by F1 and F2 formulations is shown in Table 2. F1 and F2 caused phytotoxic effects on all weeds; however, *A. hybridus* and *L. multiflorum* species were the most sensitive. F1 was more active than F2 for all weeds tested, and both formulations showed greater herbicidal activity than atrazine (positive control).

Table 2. Germination inhibition of seeds after 7 days of exposure to bioactive formulations.

Treatments	*A. hybridus*	*B. rapa*	*L. multiflorum*	*L. sativa*	*S. lycopersicum*
	IC_{50} (mM) [1] (95% Confidence Interval)				
F1	0.07 (0.03–0.09) [a]	1.5 (1.3–1.7) [a]	0.24 (0.09–0.319) [a]	2.54 (2.21–2.9) [a]	1.33 (0.89–1.63) [a]
F2	0.07 (0.04–0.08) [a]	2.3 (1.7–2.9) [b]	0.14 (0.08–0.27) [a]	4.60 (3.36–10.00) [b]	0.99 (0.37–1.38) [a]
Atrazine	41.0 (25.1–54.1) [c]	300 (189.1–368.8) [c]	38.9 (25.1–56.1) [c]	383.5 (307.1–454.8) [c]	37.5 (15.3–49.1) [c]

[1] Inhibition concentration (IC_{50}) values between treatments were considered significantly different if their confidence limits did not overlap. Different letters indicate significant differences. The experiment was performed in triplicates.

3.2.2. Effect of Binary Formulations on Photosystem II (PSII)

The effect of formulations F1 and F2 on PS II at 5 mM is shown in Figure 6. At a concentration lower than 5 mM, no significant differences among treatments were found. At 1 and 2 days after application, Fv/Fm values were lower for F1 and F2 compared to the controls. At 5 days after treatments, the Fv/Fm was zero due to the seedlings exhibiting severe damage symptoms in response to F1 and F2. However, the seedlings' exposure to atrazine showed a gradual decline in values of Fv/Fm from 5 to 12 days after treatment. In contrast, the control without herbicide did not show damage to PSII during the period evaluated (Figure 6).

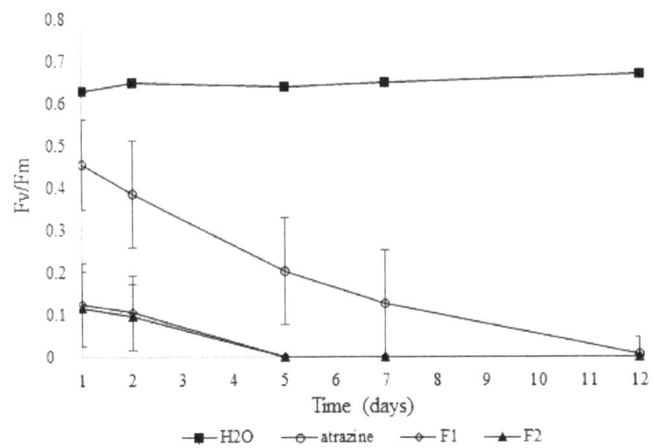

Figure 6. Effect of binary formulation F1 and F2 on PSII on seedling of *A. hybridus*.

3.2.3. Effect of Binary Formulations on Wheat Germination and Seedling Growth

The effect of F1 and F2 on wheat germination and seedling growth is shown in Figures 7 and 8. F1 and F2 were toxic 1 day after application for both concentrations. However, at 7 and 14 days after application, no significant differences were observed in the germination percentages for F1 at 2 mM and F1 at 5 mM compared to the control. Both formulations applied at 2 and 5 mM did not affect wheat germination 14 days after application (Figure 7A,B). Regarding leaf growth, both F1 and F2 at 5 mM produced growth inhibition at 7 and 14 days after application. Less growth was observed 7 days after application for both formulations at 2 mM, and no significant difference was observed 14 days after application (Figure 8A,B).

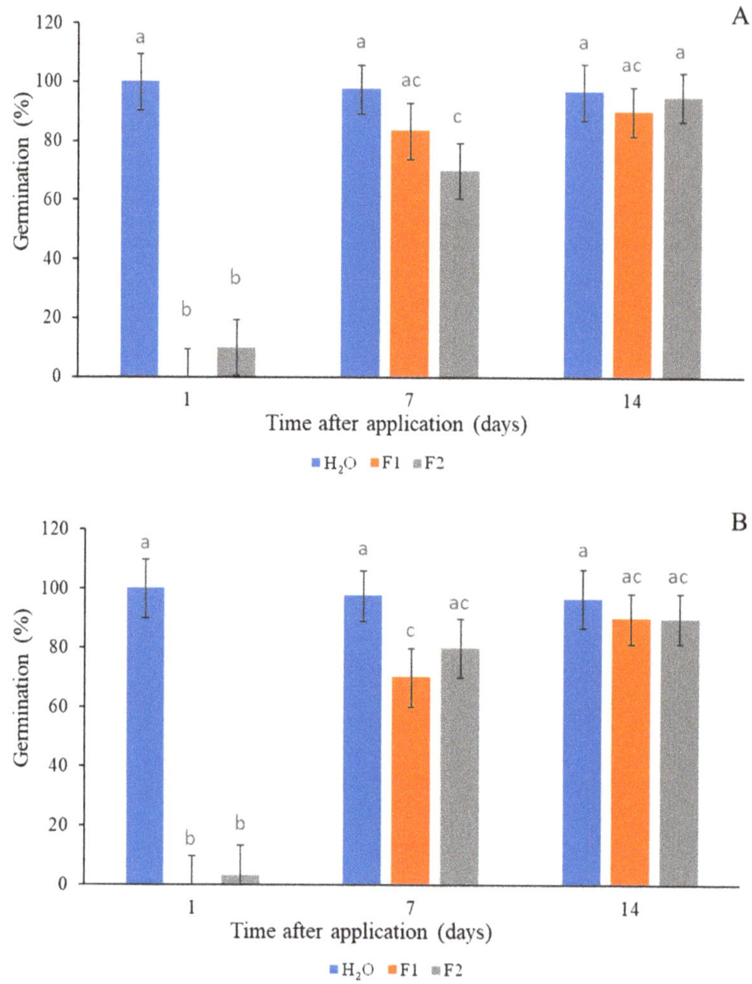

Figure 7. Percentage of germination of wheat seeds in response to different treatments (F1, orange; F2, grey and H_2O, blue) at 2 mM (**A**) and 5 mM (**B**) at 1, 7, and 14 days after application. Different letters indicate significant differences ($p < 0.05$).

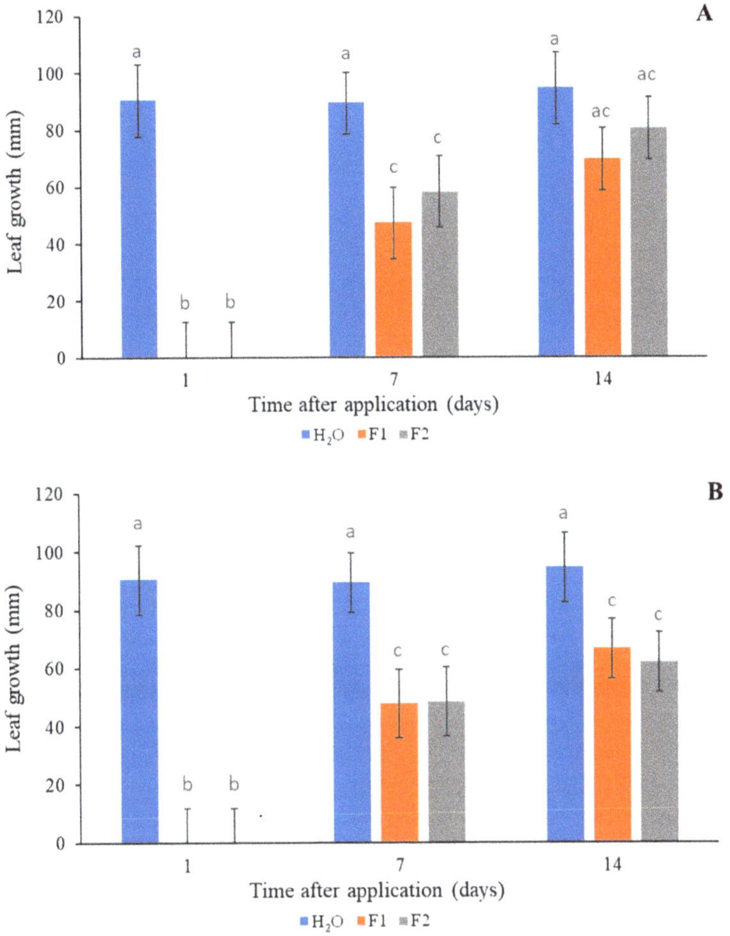

Figure 8. Leaf growth of wheat seedlings in response to different treatments (F1, orange; F2, grey and H_2O, blue) at 2 mM (**A**) and 5 mM (**B**) at 1, 7, and 14 days after application. Different letters indicate significant differences ($p < 0.05$).

4. Discussion

4.1. Chemical Characterization

The presence of active compounds and the emulsifier in herbicidal formulations (F1, F2) was confirmed by FTIR spectroscopy through the databases [32–35].

These emulsified formulations showed narrow dispersion sizes evidencing their stability between 25 and 45 °C; however, a multimodal distribution was observed for emulsifiers through DLS measurements. Current results are consistent with the complex chemical nature of soy lecithin emulsifier, which is composed of a vast mixture of amphiphilic compounds of diverse structures, mainly phospholipids and triglycerides, enabling different assembly options [32–37]. Soy lecithin has been successfully used as a natural emulsifier to improve the accessibility of different lipophilic nutrients or bioactives, showing a similar behavior [38]. Emulsification of active compounds is a key strategy in the food, cosmetic, and pharmaceutical industries [39]. The current results obtained from F1 and F2 suggest microphase reorganization into new colloidal arrangements [40]. The goal of this approach is to form stable dispersions of nonpolar additives in a polar medium (it can also be used in the opposite way) that can be properly dozed in commercial products [41].

TGA showed adequate thermal stability for nonanoic acid, citral, and thymoquinone in F1 or F2 at all temperatures tested. From these experiments, it can be observed that decomposition profiles depended on the chemical nature of active molecules; for example, citral and thymoquinone are volatile compounds and evaporate at a temperature between 45 and 165 °C, while nonanoic acid evaporates at 100 °C and above [33,42,43]. Despite ambient conditions involving the presence of atmospheric O_2, our experiments were conducted under N_2 flux. At high temperatures, oxygen may react with active compounds [44], but at temperatures tested, the oxidation reactions with small amounts of atmospheric oxygen should have a minimum impact; therefore, our results can be considered a useful approximation to actual thermal stability [33,42,43].

4.2. Herbicide Activity

The herbicide activity of F1 and F2 was evidenced on *A. hybridus*, *B. rapa*, *L. multiflorum*, *L. sativa*, and *S. lycopersicum* during the germination process, and IC50 values were from 0.07 to 4.6 mM according to the bioactive formulation and species (Table 2). F1 was more active than F2 in *B. rapa* and *L. sativa*. No differences between formulations were detected in the sensitivity of *A. hybridus*, *L. multiflorum*, and *S. lycopersicum* to both formulations. Previous studies reported by Sosa et al. [23] showed the action of plant-derived bioactive molecules on ryegrass (*L. multiflorum*) and lettuce (*L. sativa*) with IC50 of 0.6 and 0.7 mM, respectively. In agreement with our results, Dudai et al. [17] found different levels of sensitivity to citral obtained from *Cymbopogon citratus* EO (42.6% geranial and 32.1% neral) depending on the plant species (IC50 from 0.00008 to 0.000116 mM).

F1 was prepared with two bioactive agents, citral and nonanoic acid, and previous research reported that citral reduces cell division, disrupts mitotic microtubules and cell plates, and inhibits cell elongation by damaging cortical microtubules, whereas nonanoic acid causes loss of membrane integrity and rapid cell death [18,44–46]. Meanwhile, F2 consists of nonanoic acid and thymoquinone, and the mechanism of action of thymoquinone as an herbicide has not been determined. Herrera et al. [19] have reported insecticidal activity of this bioactive through acetylcholinesterase inhibition. The structure–activity research showed that topological and/or physicochemical properties can be related to the biological effects on pests. Herrera et al. [47] proposed that electronic descriptors of terpenes, such as the orbital electronegativity of the carbonyl group, are associated with enzyme inhibition in pests. Hence, F1 and F2 may be developed as promising herbicides against weeds and used in organic farming systems.

Current work demonstrates that the injuries caused by F1 and F2 on the photosynthetic apparatus of *A. hybridus* were greater than those caused by atrazine treatment (Figure 6). The measurement of Fv/Fm has been used to estimate the damage to the PSII [29,48,49]. In non-senescent and mature leaves, the values of Fv/Fm are about 0.8 [50]. For the genus *Amaranthus*, values of 0.72 were found for unstressed leaves [51]. In this study, the maximum values of Fv/Fm for the control treatment were about 0.7, and it was 0.1 1 day after the F1 or F2 treatment (Figure 6). Concerning the effects of EOs and their pure compounds on plant physiology, several studies reported a reduction in the chlorophyll content [11,18,52]. The studies of fluorescence emission analysis of photosynthetic apparatus have been used to determine the mode of action of new herbicides [53,54].

Several studies found that some EO compounds can be phytotoxic against crops [55, 56]; however, crop selectivity is a desirable trait in the development of herbicides, and the effects of F1 and F2 on wheat germination and seedling growth were approached in the current work. Both formulations were tested in pre-planting treatments, and they did not affect wheat germination when the sowing took place 7 or 14 days after application, depending on the formulation and concentration (Figure 7A,B). Regarding leaf growth, both F1 and F2 at 5 mM produced growth inhibition of wheat seedlings at 7 and 14 days after application. Less growth was observed at 7 days after application for both formulations at 2 mM, and no significant differences were observed at 14 days after application (Figure 8A,B). It is necessary to know the EO potential effect on food crops to determine the planting time at which phytotoxicity does not affect the crop. In this sense, Synowiec et al. [57] reported that caraway or peppermint EOs can selectively inhibit the growth of *Echinochloa crusgalli* but not that of maize plants. Ibañez and Blázquez [58] pointed out that oregano EO is the most harmful for cucumber and tomato seedlings, whereas rosemary EO is the least harmful for seed germination in these seedlings. This study, conducted on a laboratory scale, showed the stability and persistence of the formulations over time. We determined that after 14 days of application, the formulations did not inhibit wheat germination or growth of seedlings.

5. Conclusions

Two herbicidal formulations were developed on plant-based compounds. We demonstrated that the functional groups of the active compounds in both binary formulations were maintained, showing unimodal size distributions and thermal stability at the temperatures tested. The formulations were non-selective and controlled a broad spectrum of species (monocotyledons and dicotyledons); however, *A. hybridus* and *L. multiflorum* were the most sensitive. The binary formulations of nonanoic acid/citral and nonanoic acid/thymoquinone negatively affected the photosynthetic system of *A. hybridus*. The phytotoxic effect of the formulations was tested on wheat germination, and they did not inhibit plant germination and seedling growth after 14 days of application. However, further research should consider the validation of the effectiveness of F1 and F2 at the field scale. New herbicide formulations developed with different bioactive ingredients would control a wide spectrum of weeds through multitarget-site effects, new modes of action, and a low impact on the environment.

Author Contributions: J.J.R., J.S.-A. and J.M.H. conducted experiments, performed statistical data analysis, and wrote the manuscript. D.B.M. conducted experiments. M.E.Y. provided plant material contributed ideas and corrected the manuscript. M.E.C., S.R. and A.S.-T. performed statistical data analysis and contributed ideas. J.M.H. was involved in research design. J.J.R. and J.S.-A. contributed equally. All authors have read and agreed to the published version of the manuscript.

Funding: Financial support was provided by Y-TEC (P1635) (YPF Tecnología).

Data Availability Statement: Data are available from the authors upon reasonable request. The data are not publicly available due to privacy.

Acknowledgments: We highly appreciate the technical assistance provided by Irvicelli, Karina Giselle (Y-TEC); Rojas, Graciela Susana (Y-TEC); and Loyza; Inés de los Angeles (Y-TEC), for FTIR analyses. We thank Ana Paula Moretti (INFIVE-CONICET-UNLP), Diego Fanello (INFIVE-CONICET-UNLP), Carlos Bartoli (INFIVE-CONICET-UNLP), and Juan José Guiamet (INFIVE-CONICET-UNLP) for technical support in the study of fluorescence emission analysis. We thank Graciela Brelles-Mariño for editing the manuscript. A.S.-T., J.J.R., J.S.-A., M.E.Y. and J.M.H. are CONICET researchers.

Conflicts of Interest: Juan J. Romero, Juliana Soler-Arango, Marcos E. Coustet, Daniela B. Moracci, Sebastián Reinoso, Marcos E. Yanniccari, Aline Schneider-Teixeira are employed by the company YPF Tecnología. The last author declares no conflicts of interest. The funder was not involved in the study design, collection, analysis, interpretation of data, the writing of this article or the decision to submit it for publication.

References

1. IPCC. Intergovernmental Panel on Climate Change. Cambio Climático 2021: Bases Físicas. Contribución del Grupo de Trabajo I al Sexto Informe de Evaluación del Grupo Intergubernamental de Expertos Sobre el Cambio Climático. Available online: http://www.ipcc.ch (accessed on 7 March 2023).
2. Oreja, F.; Moreno, N.; Gundel, P.; Vercellino, R.; Pandolfo, C.; Presotto, A.; Perotti, V.; Permingeat, H.; Tuesca, D.; Scursoni, J.; et al. Herbicide-resistant weeds from dryland agriculture in Argentina. *Weed Res.* **2024**, *64*, 88–106. [CrossRef]
3. Yanniccari, M.E.; Istilart, C.; Giménez, D.; Castro, A. Glyphosate resistance in perennial ryegrass (*Lolium perenne* L.) from Argentina. *Crop. Prot.* **2012**, *32*, 12–16. [CrossRef]
4. Yanniccari, M.E.; Gaines, T.; Scursoni, J.A.; De Prado, R.; Vila-Aiub, M. Global patterns of herbicide resistance evolution in *Amaranthus spp.*: An analysis comparing species, cropping regions and herbicides. *Adv. Weed Sci.* **2022**, *40*, e0202200037. [CrossRef]
5. Víctor, J.; Núñez Fré, F.; Saint-André, H.; Fernández, R. Responses of 2,4-D resistant *Brassica rapa* L. biotype to various 2,4-D formulations and other auxinic herbicides. *Crop. Prot.* **2021**, *145*, 105621.
6. Depetri, M.; Muñiz Padilla, E.; Ayala, F.; Tuesca, D.; Breccia, G. Resistance to acetyl-CoA carboxylase (ACCase)-inhibiting herbicides in *Lolium multiflorum* Lam. populations of Argentina. *Pest Manag. Sci.* **2024**, *80*, 6600–6606. [CrossRef]
7. Ferraro, D.O.; Ghersa, F.; de Paula, R.; Duarte Vera, A.C.; Pessah, S. Historical trends of the ecotoxicological pesticide risk from the main grain crops in Rolling Pampa (Argentina). *PLoS ONE* **2020**, *15*, e0238676. [CrossRef]
8. Hasan, M.; Ahmad-Hamdani, M.S.; Rosli, A.M.; Hamdan, H. Bioherbicides: An eco-friendly tool for sustainable weed management. *Plants* **2021**, *10*, 1212. [CrossRef]
9. EPA. Available online: https://www.regulations.gov/document/EPA-HQ-OPP-2013-0266-1625 (accessed on 7 March 2023).
10. Cotrut, R. Allelopathy and allelochemical interactions among plants. *Sci. Pap.* **2018**, *1*, 188–193.
11. Verdeguer, M.; Castañeda, L.G.; Torres-Pagan, N.; Llorens-Molina, J.A.; Carrubba, A. Control of *Erigeron bonariensis* with *Thymbra capitata*, *Mentha piperita*, *Eucalyptus camaldulensis*, and *Santolina chamaecyparissus* essential oils. *Molecules* **2020**, *25*, 562. [CrossRef]
12. Kordali, S.; Cakir, A.; Ozer, H.; Cakmakci, R.; Kesdek, M.; Mete, E. Antifungal, phytotoxic and insecticidal properties of essential oil isolated from Turkish *Origanum acutidens* and its three components, carvacrol, thymol and p-cymene. *Bioresour. Technol.* **2008**, *99*, 8788–8795. [CrossRef]
13. Azirak, S.; Karaman, S. Allelopathic effect of some essential oils and components on germination of weed species. *Acta Agric. Scand. Sect. B-Soil Plant Sci.* **2008**, *58*, 88–92. [CrossRef]
14. EFSA. Conclusion on the peer review of the pesticide risk assessment of the active substance Fatty acids C7 to C18 (approved under Regulation (EC) No 1107/2009 as Fatty acids C7 to C20). *EFSA J.* **2013**, *11*, 3023. [CrossRef]
15. Devi, R.C.; Sim, S.M.; Ismail, R. Spasmolytic effect of citral and extracts of *Cymbopogon citratus* on isolated rabbit ileum. *J. Smooth Muscle Res.* **2011**, *47*, 143–156. [CrossRef]
16. Sousa, D.G.; Sousa, S.D.; Silva, R.E.; Silva-Alves, K.S.; Ferreira-da-Silva, F.W.; Kerntopf, M.R.; Menezes, I.R.; Leal-Cardoso, J.H.; Barbosa, R. Essential oil of *Lippia alba* and its main constituent citral block the excitability of rat sciatic nerves. *Braz. J. Med. Biol. Res.* **2015**, *48*, 697–702. [CrossRef] [PubMed]
17. Dudai, N.; Poljakoff-Mayber, A.; Mayer, A.; Putievsky, E.; Lerner, H. Essential oils as allelochemicals and their potential use as bioherbicides. *J. Chem. Ecol.* **1999**, *25*, 1079–1089. [CrossRef]
18. Verdeguer, M.; Sánchez-Moreiras, A.M.; Araniti, F. Phytotoxic effects and mechanism of action of essential oils and terpenoids. *Plants* **2020**, *9*, 1571. [CrossRef]
19. Herrera, J.M.; Goñi, M.L.; Gañan, N.A.; Zygadlo, J.A. An insecticide formulation of terpene ketones against *Sitophilus zeamais* and its incorporation into low density polyethylene films. *Crop. Prot.* **2017**, *98*, 33–39. [CrossRef]
20. Isaev, N.K.; Chetverikov, N.S.; Stelmashook, E.V.; Genrikhs, E.E.; Khaspekov, L.G.; Illarioshkin, S.N. Thymoquinone as a potential neuroprotector in acute and chronic forms of cerebral pathology. *Biochemistry* **2020**, *85*, 167–176. [CrossRef]
21. Hyldgaard, M.; Mygind, T.; Meyer, R.L. Essential oils in food preservation: Mode of action, synergies, and interactions with food matrix components. *Front. Microbiol.* **2012**, *3*, 12. [CrossRef]
22. Chemspider, Chemical Structure Database. Available online: http://www.chemspider.com/ (accessed on 15 January 2024).
23. Sosa, G.M.; Travaini, L.M.; Walter, H.; Cantrell, C.; Duke, S.; Carrillo, N.; Ceccarelli, E. Herbicidal Composition Comprising Chromone Derivatives and a Method for Weed Control. U.S. Patent 11083197B2, 12 December 2021.
24. YPF Agro. Available online: https://agro.ypf.com (accessed on 12 June 2024).
25. Calvimonte, H.; Peschiutta, M.L.; Herrera, J.M.; Zunino, M.P.; Jacquat, A.; Usseglio, V.L.; Zygadlo, J. Allylic and non-allylic alcohols against the maize weevil (*Sitophilus zeamais*): A promising tool for its control. *Agric. Res.* **2022**, *12*, 94–103. [CrossRef]
26. Mallory-Smith, C.A.; Retzinger, E.J. Revised classification of herbicides by site of action for weed resistance management strategies. *Weed Technol.* **2003**, *17*, 605–619. [CrossRef]
27. *Software L-POLO-Plus a User's Guide to Probit or Logic Analysis*; LeOra Software: Berkeley, CA, USA, 2002.

28. Busi, R.; Goggin, D.E.; Heap, I.M.; Horak, M.J.; Jugulam, M.; Masters, R.A.; Napier, R.M.; Riar, D.S.; Satchivi, N.M.; Torra, J.; et al. Weed resistance to synthetic auxin herbicides. *Pest Manag. Sci.* **2018**, *74*, 2265–2276. [CrossRef] [PubMed]
29. Pooja, M.; Anjana, J.; Sonal, M.; Sudhakar, B. Chlorophyll a fluorescence study revealing effects of high salt stress on Photosystem II in wheat leaves. *Plant Physiol. Biochem.* **2010**, *48*, 16–20.
30. Di Rienzo, J.A.; Casanoves, F.; Balzarini, M.G.; Gonzalez, L.; Tablada, M.; Robledo, C.W. *InfoStat*; versión 2017; Universidad Nacional de Córdoba: Córdoba, Argentina, 2017.
31. Peschiutta, M.L.; Brito, V.D.; Achimon, F.; Zunino, M.P.; Usseglio, V.L.; Zygadlo, J. New insecticide delivery method for the control of *Sitophilus zeamais* in stored maize. *J. Stored Prod. Res.* **2019**, *83*, 185–190. [CrossRef]
32. Van Nieuwenhuyzen, W.; Szuhaj, B.F. Effects of lecithins and proteins on the stability of emulsions. *Lipid/Fett* **1998**, *100*, 282–291. [CrossRef]
33. Wang, Z.; Sun, J.; Xie, S.; Ma, G.; Jia, Y. Thermal properties and reliability of a lauric acid/nonanoic acid binary mixture as a phase-change material for thermal energy storage. *Energy Technol.* **2017**, *5*, 2309–2316. [CrossRef]
34. Whittinghill, J.M.; Norton, J.; Proctor, A. Stability determination of soy lecithin-based emulsions by Fourier Transform Infrared Spectroscopy. *J. Am. Oil Chem. Soc.* **2000**, *77*, 37–42. [CrossRef]
35. Herrera, J.M.; Peralta, E.; Palacio, M.A.; Mercado Ruiz, J.N.; Strumia, M.C.; Zygadlo, J.A.; Soto Valdez, H. Bioactive silo bag for controlling stored pest: *Sithophilus zeamais* (Coleoptera: Curculionidae). *Rev. Ibero. Tec. Post.* **2022**, *23*, 214–223.
36. Scholfield, C.R. Composition of soybean lecithin. *J. Am. Oil Chem. Soc.* **1981**, *58*, 889–890. [CrossRef]
37. Taladrid, D.; Marín, D.; Alemán, A.; Álvarez-Acero, I.; Montero, P.; Gómez-Guillén, M.C. Effect of chemical composition and sonication procedure on properties of food-grade soy lecithin liposomes with added glycerol. *Food Res. Int.* **2017**, *100*, 541–550. [CrossRef]
38. Yang, Q.Q.; Sui, Z.; Lu, W.; Corke, H. Soybean lecithin-stabilized oil-in-water (O/W) emulsions increase the stability and in vitro bioaccessibility of bioactive nutrients. *Food Chem.* **2021**, *338*, 128071. [CrossRef]
39. Chung, C.; Sher, A.; Rousset, P.; Decker, E.A.; McClements, D.J. Formulation of food emulsions using natural emulsifiers: Utilization of quillaja saponin and soy lecithin to fabricate liquid coffee whiteners. *J. Food Eng.* **2017**, *209*, 1–11. [CrossRef]
40. Zaldivar, G.; Perez Sirkin, Y.A.; Debais, G.; Fiora, M.; Missoni, L.L.; Gonzalez Solveyra, E.; Tagliazucchi, M. Molecular theory: A tool for predicting the outcome of self-assembly of polymers, manoparticles, amphiphiles, and other soft materials. *ACS Omega* **2022**, *7*, 38109–38121. [CrossRef] [PubMed]
41. Grillo, I.; Morfin, I.; Prévost, S. Structural characterization of Pluronic micelles swollen with perfume molecules. *Langmuir* **2018**, *34*, 13395–13408. [CrossRef] [PubMed]
42. Pagola, S.; Benavente, A.; Raschi, A.R.; Molina, M.A.A.; Stephens, P.W. Crystal structure determination of thymoquinone by high-resolution X-ray powder diffraction. *Aaps Pharmscitech* **2004**, *5*, 24–31. [CrossRef] [PubMed]
43. Aytac, Z.; Celebioglu, A.; Yildiz, Z.I.; Uyar, T. Efficient encapsulation of citral in fast-dissolving polymer-free electrospun nanofibers of cyclodextrin inclusion complexes: High thermal stability, longer shelf-life, and enhanced water solubility of citral. *Nanomaterials* **2018**, *8*, 793. [CrossRef]
44. Senseman, S.A. *Herbicide Handbook*, 9th ed.; Weed Science Society of America: Champaign, IL, USA, 2007; 458p.
45. Chaimovitsh, D.; Rogovoy, O.; Altshuler, O.; Belausov, E.; Abu-Abied, M.; Rubin, B.; Sadot, E.; Dudai, N. The relative effect of citral on mitotic microtubules in wheat roots and BY2 cells. *Plant Biol.* **2012**, *14*, 354–364. [CrossRef]
46. Graña Martinez, E. Mode of Action and Herbicide Potential of the Terpenoids Farnesene and Citral on "Arabidopsis thaliana" Metabolism. 2015. Available online: https://dialnet.unirioja.es/servlet/tesis?codigo=124599 (accessed on 12 June 2024).
47. Herrera, J.M.; Zygadlo, J.A.; Strumia, M.C.; Peralta, E. Biopesticidal silo bag prepared by co-extrusion process. *Food Packag. Shelf Life* **2021**, *28*, 100645. [CrossRef]
48. Govindjee. Chlorophyll a fluorescence: A bit of basics and history. In *Chlorophyll a Fluorescence. Advances in Photosynthesis and Respiration*; Papageorgiou, G.C., Govindjee, Eds.; Springer: Dordrecht, The Netherlands, 2004; Volume 19, pp. 1–48.
49. Slabbert, R.M.; Krüger, G.H. Assessment of changes in photosystem II structure and function as affected by water deficit in *Amaranthus hypochondriacus* L. and *Amaranthus hybridus* L. *Plant Physiol. Biochem.* **2011**, *49*, 978–984. [CrossRef]
50. Pino, M.T. *Estrés hídrico y térmico en papas, avances y protocolos. Santiago, Chile. Instituto de Investigaciones Agropecuarias*; Instituto de Investigaciones Agropecuarias: Santiago, Chile, 2016; Boletín INIA N ° 331; p. 148. Available online: https://biblioteca.inia.cl/items/7cbd1f1a-aeb4-47a9-b3a5-e432821ad95f (accessed on 12 June 2024).
51. Cai, Y.Z.; Corke, H.; Wu, H.X. Amaranth. In *Encyclopedia of Grain Science*; The University of Hong Kong: Hong Kong, 2004; pp. 1–10. [CrossRef]
52. Verdeguer, M.; Blázquez, M.A.; Boira, H. Phytotoxic effects of *Lantana camara*, *Eucalyptus camaldulensis* and *Eriocephalus africanus* essential oils in weeds of Mediterranean summer crops. *Biochem. Syst. Ecol.* **2009**, *37*, 362–369. [CrossRef]
53. Maxwell, K.; Johnson, G.N. Chlorophyll fluorescence—A practical guide. *J. Exp. Bot.* **2000**, *51*, 659–668. [CrossRef] [PubMed]
54. González Moreno, S.; Perales, V.; Salcedo Alvarez, H.; Martha, O. La fluorescencia de la clorofila a como herramienta en la investigación de efectos tóxicos en el aparato fotosintético de plantas y algas. *Rev. Educ. Bioquímica* **2008**, *27*, 119–129.

55. Amri, I.; Hamrouni, L.; Hanana, M.; Jamoussi, B. Reviews on phytotoxic effects of essential oils and their individual components: News approach for weed management. *Int. J. Appl. Biol. Pharm. Technol.* **2013**, *4*, 96–114.
56. Werrie, P.Y.; Durenne, B.; Delaplace, P.; Fauconnier, M.L. Phytotoxicity of essential oils: Opportunities and constraints for the development of biopesticides. *A Review Foods* **2020**, *9*, 1291. [CrossRef]
57. Synowiec, A.; Możdżeń, K.; Krajewska, A.; Landi, M.; Araniti, F. Carum carvi L. essential oil: A promising candidate for botanical herbicide against *Echinochloa crus-galli* (L.) P. Beauv. in maize cultivation. *Ind. Crops Prod.* **2019**, *140*, 111652. [CrossRef]
58. Ibáñez, M.D.; Blázquez, M.A. Phytotoxic effects of commercial essential oils on selected vegetable crops: Cucumber and tomato. *Sustain. Chem. Pharm.* **2020**, *15*, 100209. [CrossRef]

plants

MDPI

Article

Effects of Submerged Macrophytes on Demography and Filtration Rates of *Daphnia* and *Simocephalus* (Crustacea: Cladocera)

Cristian A. Espinosa-Rodríguez [1], Alfonso Lugo-Vázquez [1], Luz J. Montes-Campos [1], Ivan M. Saavedra-Martínez [1], Ma. del Rosario Sánchez-Rodríguez [1], Laura Peralta-Soriano [1] and Ligia Rivera-De la Parra [2,*]

[1] Grupo de Investigación en Limnología Tropical, UIICSE, FES Iztacala, Universidad Nacional Autónoma de México, Av. De los Barrios 1, Col. Los Reyes Iztacala, Tlalnepantla CP 54090, Estado de México, Mexico; caer_atl@iztacala.unam.mx (C.A.E.-R.); lugov@unam.mx (A.L.-V.); luzjazmin58@gmail.com (L.J.M.-C.); ivansaavedrarobot@gmail.com (I.M.S.-M.); rosarios@unam.mx (M.d.R.S.-R.); sorial@unam.mx (L.P.-S.)

[2] Laboratorio de Fisiología Vegetal, L-204, FES Iztacala, Universidad Nacional Autónoma de México, Av. De los Barrios 1, Col. Los Reyes Iztacala, Tlalnepantla CP 54090, Estado de México, Mexico

* Correspondence: ligia.lrp@gmail.com

Abstract: Macrophytes and cladocerans represent the main antagonistic groups that regulate phytoplankton biomass; however, the mechanism behind this interaction is unclear. In laboratory conditions, we separately evaluated the effects of three submerged macrophytes (*Ceratophyllum demersum*, *Myriophyllum aquaticum*, and *Stuckenia pectinata*), as well as their exudates, and plant-associated microbiota (POM < 25 μm) + exudates on the population growth of *Daphnia* cf. *pulex* and *Simocephalus* cf. *mixtus*. Living *Ceratophyllum*, exudates, and POM < 25 μm + exudates exhibited the most robust positive effects on *Simocephalus* density and the rate of population increase (r). Subsequently, we examined the effects of *Ceratophyllum* on the filtration and feeding rates of *Simocephalus* and *Daphnia*, revealing significant ($p < 0.001$) promotion of filtration and feeding in *Simocephalus* but not in *Daphnia*. To elucidate the specific effects of this macrophyte on *Simocephalus* demography, we assessed selected life table variables across the same treatments. The treatments involving exudates and living *Ceratophyllum* resulted in approximately 40% longer survivorship and significantly ($p < 0.01$) enhanced fecundity. Our findings indicate that exudates from submerged macrophytes positively influence *Simocephalus* demography by increasing filtration rates, survivorship, and fecundity. This synergy suggests a substantial impact on phytoplankton abundance.

Keywords: exudates; feeding rates; life table; population growth rate; zooplankton

1. Introduction

During recent decades, a significant increase in urbanization, agriculture, livestock, deforestation, and wastewater discharges has been observed. These activities have expedited the eutrophication process in numerous water bodies around the world, leading to substantial alterations in biological communities and ecosystem dynamics [1]. Due to this environmental deterioration and global warming, eutrophication is one of the main problems facing marine and freshwater ecosystems [2]. This phenomenon drives an increase in phytoplankton biomass with associated toxic potential and reduces aquatic diversity, diminishing ecosystem services' availability [3,4]. For this reason, several methods have been proposed to mitigate this problem. These include physical methods such as algae harvesting, dredging, water diversion, shading of lake areas, and the use of ultrasonic waves to disrupt algae cells. Chemical methods include the application of substances that kills algae, flocculants, and growth regulators [5–7]. However, many of these are ineffective, have harmful side effects, or are too expensive to implement [8]. In this sense, biomanipulation is a restoration method mainly applied in lakes and reservoirs, allowing for improving water quality in short periods at relatively low costs [3,8].

Submerged macrophytes have been widely used in ecological restoration projects since they are a key element in shaping aquatic communities [3,9,10]. They establish antagonistic interactions with phytoplankton by enhancing sedimentation rates, engaging in competition for light and nutrients, providing refuge for herbivores [11], and producing allelopathic compounds that reduce phytoplankton abundance [12]. Several works have shown that the allelopathic potential is more remarkable in certain species of submerged macrophytes such as *Myriophyllum* spp., *Chara* spp., *Potamogeton* spp., and *Ceratophyllum demersum* [13], with different kinds of identified compounds as major chemical mediators of this phenomenon [12,13]. These substances are released and diffused as dissolved organic carbon (DOC) into the environment, rapidly degraded by heterotrophic bacteria and thereby stimulating the microbial food web [14]. However, there is not much information at an experimental level about how these processes occur and their effect on higher trophic levels, as is the case of primary consumers such as cladocerans [15].

This interaction between macrophytes and cladocerans is relevant in biomanipulation projects since macrophytes also help to reduce phytoplankton biomass by providing shelter for zooplankton against predation, which increases herbivory rates [9]. However, the outcome of the macrophyte–cladoceran interaction is unclear since different responses have been reported in cladocerans at a demographic, physiological, and behavioral level [16–19]. Furthermore, most of these studies have been carried out in temperate systems, and several works have shown that ecological interactions in tropical and subtropical systems may differ from those observed in temperate systems [20]. Burns and Dodds [16] documented contrasting impacts of *Nitella hookeri* exudates on the filtration rates of *Daphnia carinata*, revealing both positive and negative effects depending on seasonal variations. Burks et al. [17] found that *Elodea canadensis* exudates caused a delay in maturation time and decreased egg production in *Daphnia magna*. Later, Cerbin et al. [18] mentioned that *Myriophyllum verticillatum* induced a reduction in first reproduction size and clutch size in *D. magna*, although no discernible adverse effects were observed from its exudates. Repellence effects from cladocerans towards macrophytes have also been documented [19,21]. Conversely, Espinosa-Rodríguez et al. [22] observed an enhancement in the longevity and fecundity of three *Simocephalus* species following exposure to exudates from *Egeria densa*. Montiel-Martínez [23] indicated that exudates from the floating macrophyte *Eichhornia crassipes* influenced behavior and caused positive effects on demographic parameters of littoral cladocerans, specifically *Simocephalus vetulus* and *Chydorus brevilabris*.

According to this, macrophytes may exert negative effects on pelagic cladocerans such as *Daphnia*, whereas littoral cladocerans that coevolve with macrophytes, such as those belonging to the *Simocephalus* genus, may experience positive effects. However, empirical evidence supporting these hypotheses is lacking. Consequently, the impact of submerged macrophytes, their exudates, and associated microbiota on the demographic characteristics and filtration rates of littoral and pelagic cladocerans remains uncertain. The hypothesized mechanism that explains the negative effect of macrophytes on cladocerans is associated with the release of allelopathic substances [9]; nevertheless, the chemical composition of these substances varies among macrophyte species, and their effects depend on the recipient species' sensitivity [12,13]. On the other hand, the hypothesis suggesting a positive impact of macrophyte exudates on cladocerans relies on the fact that these substances, in the form of dissolved organic carbon (DOC), are readily decomposed by heterotrophic bacteria, which utilize them as an energy source [24]. This decomposition process activates the microbial food web, which serves as a food source for zooplankton [25], so the nature and quantity of these substances will influence bacterial production, and this will depend on the macrophyte species involved and prevailing environmental conditions [14,26]. In this study, we aimed to assess (a) the effect of three submerged macrophytes (*Ceratophyllum demersum*, *Myriophyllum aquaticum*, and *Stuckenia pectinata*), as well as their exudates and associated particulate organic matter smaller than 25 μm (POM > 25 μm), on the population growth of *Daphnia* cf. *pulex* and *Simocephalus* cf. *mixtus*; (b) the influence of *C. demersum* and *S. pectinata* on the feeding and filtration rates of *D.* cf. *pulex* and *S.* cf. *mixtus*; and (c) the effects of *C. demersum* on the survivorship and fecundity of *S.* cf. *mixtus*.

2. Materials and Methods

2.1. Test Organisms

The plankton species used for the tests (*Chlamydomonas* sp., *Daphnia* cf. *pulex*, and *Simocephalus* cf. *mixtus*) were isolated from the urban Lake Mexcalpique within Cantera Oriente. For *Chlamydomonas* sp., we directly collected the sample from the surface layer using a 50 mL Falcon tube; cladocerans were qualitatively sampled from the same layer with a zooplankton net of 64 µm mesh size. La Cantera Oriente is a place situated in the protected natural area known as "Reserva del Pedregal de San Ángel" (REPSA), Mexico City, with the coordinates ranging between 19°18′47″ and 19°19′15″ N and between 99°10′17″ and 99°10′22″ W. This region encompasses a spring and 4 shallow lakes exhibiting distinct trophic levels; therefore, it could be used as a model for assessing lake restoration strategies [27]. In these lakes, *Chlamydomonas* spp. blooms have been recorded during the cold season (November to March), while the chosen cladocerans represent the largest filter-feeding species present in these lakes. The selected aquatic plant species are distributed across several water bodies in central Mexico. *Myriophyllum aquaticum* and *Stuckenia pectinata* were isolated from the Salazar reservoir in the State of Mexico, whereas *Ceratophyllum demersum* was isolated from the channels of Xochimilco in Mexico City.

2.2. Culture and Maintenance of Organisms

Chlamydomonas sp. was isolated in 1.5% bacteriological agar Petri dishes with Bold basal medium as per the method described by Andersen et al. [28]. It was then transferred to a liquid Bold basal medium for scaling up and cultured in 500 mL transparent glass bottles at a controlled temperature of 21 ± 2 °C under aeration and constant, diffuse fluorescent light conditions. Following an incubation period of 8 days, the algae were harvested, centrifuged at 3000 RPM for 5 minutes, and resuspended in EPA medium [29]. Algal concentrations were determined using a hematocytometer. Cladoceran cultures were initiated from parthenogenetic females and maintained for 8 months in EPA medium prepared with 96 mg of $NaHCO_3$, 60 mg of $CaSO_4$, 60 mg of $MgSO_4$, and 4 mg of KCl in 1 L of deionized water [29], at a temperature of 21 ± 2 °C under a 12:12 photoperiod, with *Chlamydomonas* sp. (at approximately 1×10^6 cells mL^{-1}) provided as the food source. The culture medium was replaced twice weekly. Macrophytes were collected from the field and underwent a week-long quarantine period with daily dechlorinated tap water changes to minimize associated biota. Following quarantine, they were placed in a 10% Bold basal medium under aeration, at the same temperature and photoperiod as the other organisms. To prepare the exudates and particulate organic matter less than 25 µm (POM < 25 µm), each macrophyte was rinsed with EPA medium and placed in a 5 L aquarium at a density of 1.2 g dry weight L^{-1} in 4 L of EPA medium + 10% Bold basal medium 48 h before the start of the experiment. Aquarium conditions, including light, photoperiod, and aeration, were maintained identical to those for the algae to ensure an oxygenic environment and prevent gradients. Water levels were replenished daily with EPA medium + 10% Bold basal medium.

2.3. Experimental Design

Four treatments were established for each macrophyte and they were evaluated separately for each cladoceran species. The treatments were (1) macrophyte exudates, (2) macrophyte exudates + associated microbiota and detritus less than 25 µm (POM < 25 µm), (3) living macrophytes at a density of 1.2 g dry weight L^{-1}, and (4) control. All treatments had basal Bold medium 10% served as the nutrient source. Each treatment was replicated four times, resulting in a total of 96 experimental units (2 cladocerans × 3 macrophytes × 4 treatments × 4 replicates). For the treatments, each macrophyte was maintained as described previously. Exudates were collected by filtering the conditioned medium from each plant through 0.45 µm nitrocellulose filters. For POM < 25 µm, water containing exudates from each macrophyte was filtered through a sieve with a mesh opening size of 25 μm. In the control treatment, only EPA medium + basal Bold medium 10% was utilized.

2.4. Population Growth

Population growth experiments were conducted over 23–24 days. Each experimental unit comprised 100 mL transparent containers with 50 mL of medium with a concentration of 0.5×10^6 cells mL^{-1} of *Chlamydomonas* sp. as the primary food source. Within each container, 10 parthenogenetic females (0.2 ind. mL^{-1}) from each cladoceran species were introduced, considering cohorts consisting of 5 neonates, 3 juveniles, and 2 adults. Once the experiments began, the daily population abundance of surviving organisms was quantified, and they were subsequently transferred using a Pasteur pipette to another container corresponding to their treatment group. Based on these abundance data, population growth graphs were generated, and the daily population growth rate (r) was calculated using the Krebs exponential equation [30]:

$$r = (\ln Nt - \ln No)/t$$

where r represents the population increase rate per day, No denotes the initial population density, Nt is the final population density, and t represents the time in days. The population growth rate was calculated considering the time from the beginning of the experiment until the first point of maximum abundance in the curves for all tests. Following this calculation, growth rates were compared between treatments using a one-way ANOVA, and a post hoc Tukey test was employed using Sigmaplot 14.0 (Systat Software, Erkrath, Germany) to identify significant differences when they were observed [31].

2.5. Filtration and Feeding Rates

From the population growth experiments, we selected *C. demersum* and *S. pectinata* because these macrophytes had a more significant impact on the population growth curves of cladocerans. To assess the impact of *Ceratophyllum demersum* and *Stuckenia pectinata* exudates on the filtration and feeding rates of two cladoceran species, we established two transparent containers of 2 L filled with 1.5 liters of EPA medium at a controlled temperature of 21 ± 2 °C under a 12:12 photoperiod, with *Chlamydomonas* sp. provided as the food source, consistent with regular culture maintenance practices. *Ceratophyllum demersum* was added to one container at a density of 1.2 g dry weight L^{-1}, while the other container received *Stuckenia pectinata* at the same density. Each container was populated with a separate cohort of either *Simocephalus* cf. *mixtus* or *Daphnia* cf. *pulex*, introduced 72 h before the experiments for acclimatization. Controls were maintained under identical conditions in two separate containers for each cladoceran species but without any plants, resulting in a total of six containers.

Prior to experimentation, cladocerans were starved for 30 min, after which five adults of *D.* cf. *pulex* (2371 ± 155 μm) and *S.* cf. *mixtus* (1781 ± 163 μm) were selected for each experimental unit, represented by transparent containers with 50 mL of the respective treatment medium with a concentration of 0.5×10^6 cells mL^{-1} of *Chlamydomonas* sp. In total, 40 containers were set up (2 macrophytes × 2 cladocerans × 2 treatments × 5 replicates) with an additional 5 control containers only with *Chlamydomonas* sp. cells.

After 30 min of feeding, all experimental units were fixed with Lugol's solution 1%, and the remaining *Chlamydomonas* sp. cells were counted using a hemocytometer in a microscope Axiostar (Carl Zeiss, Jena, Germany). Filtration rates (F) and feeding rates (f) were calculated according to Rigler [32] as follows:

$$F = (\ln C0 - \ln Ct) * W/t * N$$

where C0 is the initial cell density, Ct is the final cell density, W is the medium volume in milliliters, t is the feeding time in min, and N is the number of individuals per recipient.

$$(f) = V(C0 - Ct)/t - N$$

where V is the medium volume, C0 is the initial cell density, Ct is the final cell density, t is the time, and N is the number of individuals per recipient.

Statistical comparisons were conducted independently for each cladoceran species in the presence and absence of *C. demersum* and *S. pectinata*. Thus, the results were statistically compared using one-way ANOVA and a post hoc Tukey test with Sigmaplot 14.0 (Systat Software, Erkrath, Germany) when differences were registered [31].

2.6. Life Table

For a detailed analysis of the observed effects on survivorship and fecundity registered in the population growth experiments and filtration and feeding rates, we conducted life table experiments using the same treatments outlined previously, focusing only on *Ceratophyllum demersum* and *Simocephalus* cf. *mixtus* given the pronounced effect of macrophytes on cladocerans registered in this combination. In each experimental unit, 10 neonates younger than 24 h were introduced each day; the number of survivors was quantified, with neonates being removed from the original cohort each day and transferred to another container corresponding to their treatment group. Life table parameters including survival (l_x) and fecundity (m_x) were used to calculate average lifespan (ALS), life expectancy (LE), gross (GRR) and net reproductive rate (NRR), generation time (GT), and the rate of population increase (RPI) per day following the methodology described by Krebs [30]:

Average lifespan:

$$\sum_{0}^{\infty} l_x$$

Life expectancy:

$$e_0 = \frac{T_x}{n_x}, T_x = \sum_{0}^{\infty} L_x, L_x = \frac{n_x + n_{x+1}}{2}$$

Gross reproductive rate:

$$\sum_{0}^{\infty} m_x$$

Net reproductive rate:

$$R_0 = \sum_{0}^{\infty} l_x m_x$$

Generation time:

$$T = \frac{\sum l_x m_x * x}{R_0}$$

Population growth rate (r):

$$\sum_{x=0}^{\infty} e^{-rx} l_x m_x = 1$$

where T_x is the number of individuals per day, n_x is the number of living organisms at the beginning, l_x is the probability of an organism surviving at an age class (x), m_x is the fecundity at a specific age, R_0 is the average number of neonates per female, and r is the rate of population increase. Differences between treatments were compared using one-way ANOVA, followed by a post hoc Tukey test with Sigmaplot 14.0 (Systat Software, Erkrath, Germany) to detect significant differences when they were registered [31].

3. Results

3.1. Population Growth

The population growth curves (Figure 1) showed that treatments incorporating exudates, POM, and living macrophytes of *Ceratophyllum demersum* resulted in an increase in *Simocephalus* abundance by close to 5 ind. mL^{-1}, compared to control groups that reached densities close to 2 ind. mL^{-1}. Conversely, for *Daphnia*, the presence of living *C. demersum* led to slightly higher abundances (2.37 ± 0.24 ind. mL^{-1}) compared to other treatments (POM = 2.04 ± 0.35 and exudates = 2.22 ± 0.11 ind. mL^{-1}) and the control (2.07 ± 0.33 ind. mL^{-1}); however, the control group's abundance declined earlier. Similarly, experiments with *M. aquaticum* also showed higher abundances in treatments for living macrophytes for both *Simocephalus* (1.98 ± 0.29 ind. mL^{-1}) and *Daphnia* (1.93 ± 0.14 ind. mL^{-1}) compared to other treatments, although the differences among treatments and densities were less pronounced. When *Simocephalus* was exposed to living *Stuckenia*, its population density reached 5.28 ± 0.45 ind. mL^{-1}, significantly higher than the control (1.21 ± 0.09 ind. mL^{-1}), while treatments with exudates and POM had slightly higher densities than the control, with 1.27 ± 0.14 ind. mL^{-1} and 1.52 ± 0.55 ind. mL^{-1}, respectively. In contrast, for *Daphnia*, no clear differences were observed in the abundance among all treatments, ranging between 1.92 and 2.4 ind. mL^{-1}; nevertheless, the treatment with living macrophytes showed a longer lag phase.

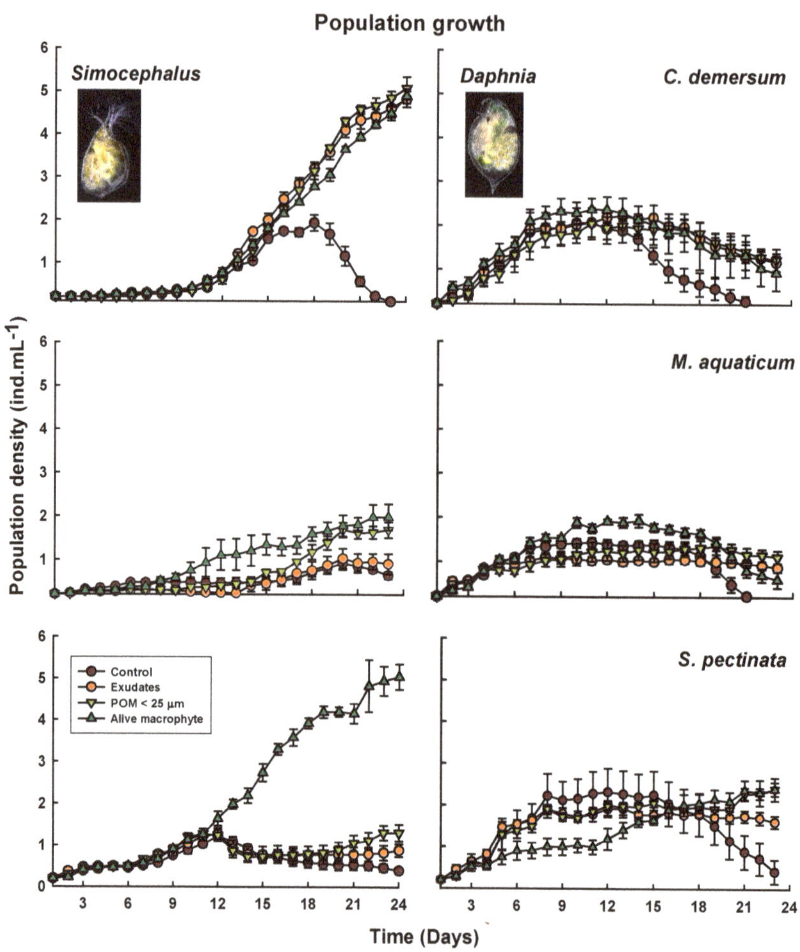

Figure 1. Population growth curves of *Simocephalus* cf. *mixtus* and *Daphnia* cf. *pulex* under the influence of exudates, particulate organic matter smaller than 25 µm (POM < 25 µm) + exudates, and living macrophytes derived from *Ceratophyllum demersum*, *Myriophyllum aquaticum*, and *Stuckenia pectinata*. Mean ± SE data are shown based on four replicates.

Overall, the r values ranged from 0.05 to 0.18 for *Simocephalus* and from 0.14 to 0.31 for *Daphnia*, as depicted in Figure 2. Regarding *Simocephalus*, all treatments involving exudates, POM, and living macrophytes yielded higher r values. However, for *Daphnia*, lower r values were generally observed in treatments with macrophytes or their derivates. In the ANOVA tests conducted with *Ceratophyllum* and *Simocephalus*, statistically significant differences ($p < 0.001$) were noted between the control group and the other treatments, with the highest r values for this cladoceran observed in the treatment involving exudates. Conversely, for *Daphnia*, no statistical differences ($p > 0.05$) were observed among the treatments. When using *M. aquaticum*, the lowest r values were observed for both cladocerans; however, only *Simocephalus* exhibited statistical differences ($p < 0.01$) between the control group and treatments with alive macrophytes. For *Daphnia*, no significant differences ($p > 0.05$) were detected. In the experiments with *Stuckenia*, statistical differences ($p < 0.001$) were observed between the control group and treatments with living macrophytes for *Simocephalus*, while for *Daphnia*, living *Stuckenia* significantly reduced ($p < 0.001$) its rate of population increase.

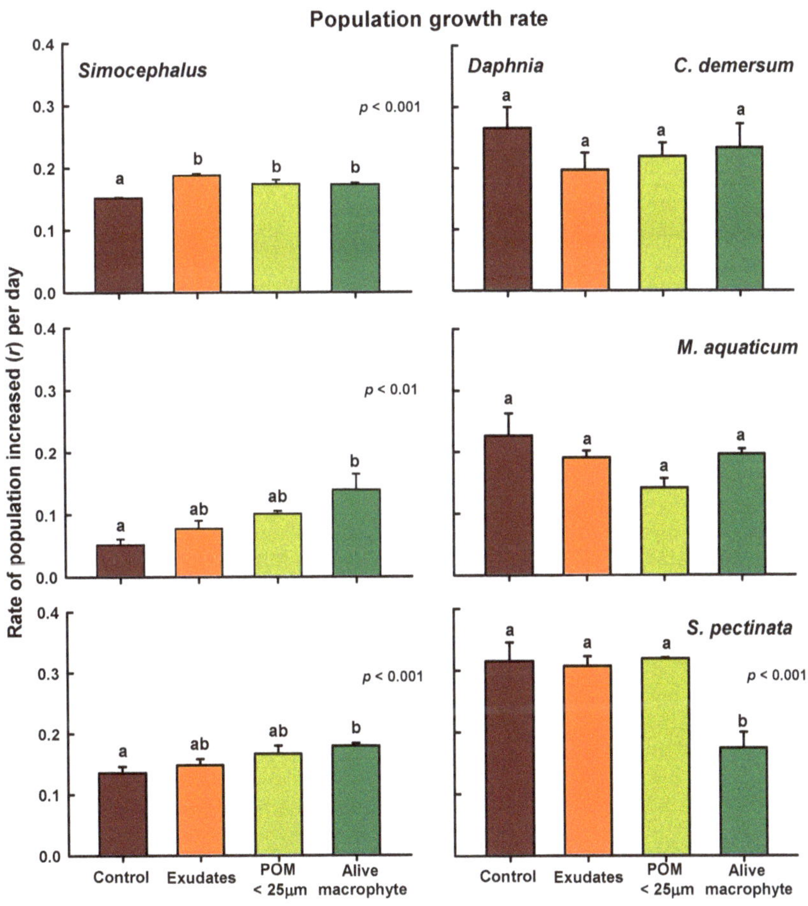

Figure 2. Population growth rates (r) per day of *Simocephalus* cf. *mixtus* and *Daphnia* cf. *pulex* under the influence of exudates, particulate organic matter smaller than 25 μm (POM < 25 μm) + exudates, and living macrophytes derived from *Ceratophyllum demersum*, *Myriophyllum aquaticum*, and *Stuckenia pectinata*. Mean ± SE data are shown based on four replicates. Bars with equal letters are not statistically significant ($p > 0.05$, Tukey test).

3.2. Filtration and Feeding Rates

The filtration and feeding rate values for *Simocephalus* significantly increased ($p < 0.001$) in the presence of *C. demersum* and *S. pectinata*, while *Daphnia* did not show statistical differences ($p > 0.05$; Table 1).

Table 1. Filtration and feeding rates of *Simocephalus* cf. *mixtus* and *Daphnia* cf. *pulex* in the presence and absence of *Ceratophyllum demersum* and *Stuckenia pectinata*. Means ± SE are based on five replicates. Data with equal letters are not statistically significant ($p > 0.05$, Tukey test).

		Simocephalus cf. *mixtus*	*Daphnia* cf. *pulex*
Filtration rates (mL ind. h^{-1})	Control	1.01 ± 0.19 ^{a}	2.14 ± 0.24 ^{a}
	C. demersum	4.07 ± 0.52 ^{b}	1.99 ± 0.32 ^{a}
	S. pectinata	3.02 ± 0.30 ^{c}	1.88 ± 0.15 ^{a}
Feeding rates $\times 10^5$ (cells ind. h^{-1})	Control	0.81 ± 0.15 ^{a}	1.63 ± 0.17 ^{a}
	C. demersum	2.87 ± 0.31 ^{b}	1.52 ± 0.22 ^{a}
	S. pectinata	2.22 ± 0.20 ^{c}	1.45 ± 0.25 ^{a}

3.3. Life Table

Life table experiments showed that the survivorship (l_x) of *Simocephalus* notably increased in the presence of exudates and living *C. demersum*, with individuals living up to 54 and 52 days, respectively, compared to 38 days in the POM < 25 μm treatment, and 33 days in the control group. The maximum fecundity (m_x) values were relatively similar between the control and POM treatments, averaging 3.5 and 3.4 neonates per female^{-1}, respectively, and cumulated fecundity around 30 neonates per female^{-1}. In contrast, treatments with exudates and living macrophytes exhibited higher values of 4.6 and 6.33 neonates per female^{-1} as well as a cumulated fecundity of 36 and 52 neonates per female^{-1}, respectively (Figure 3).

Table 2 displays selected life table variables for *Simocephalus*, comparing exudates, POM, and alive *C. demersum*. The average life span was similar ($p > 0.05$) between control and POM treatments, while exudates and alive *C. demersum* showed statistically different values ($p < 0.01$ and $p < 0.001$ respectively) compared to the control group. Both gross and net reproductive rates exhibited a similar trend, with similar lower values in the control group compared to the other treatments; however, exudates and POM showed no significant difference ($p > 0.05$). Notably, alive *C. demersum* resulted in statistically higher values ($p < 0.001$) compared to the other treatments. Furthermore, generation time was significantly higher for alive *C. demersum* ($p < 0.01$) and lower for POM; although, the latter was similar to the control group ($p > 0.05$). Concerning the rate of population increase, treatments with exudates and alive *Ceratophyllum* were significantly higher ($p < 0.01$ and $p < 0.05$ respectively) compared to the control group, while POM showed no statistical differences ($p > 0.05$).

Table 2. Selected life table variables of *Simocephalus* cf. *mixtus* under the influence of exudates, particulate organic matter smaller than 25 μm + exudates (POM < 25 μm), and living *Ceratophyllum demersum*. Mean ± SE data are shown based on four replicates. Data with equal letters are not statistically significant ($p > 0.05$, Tukey test). ALS = Average life span; GRR = gross reproductive rate; NRR = net reproductive rate; GT = generational time; RPI = rate of population increase.

Treatments	Variable	*Simocephalus*
Control	ALS (days)	26.6 ± 0.33 ^{a}
Exudates		37.9 ± 2.45 ^{b}
POM < 25 μm		28.7 ± 0.87 ^{a}
Alive *C. demersum*		39.0 ± 1.43 ^{b}

Table 2. *Cont.*

Treatments	Variable	*Simocephalus*
Control	GRR (neonates per female^{-1})	23.3 ± 1.23 [a]
Exudates		37.1 ± 2.64 [b]
POM < 25 µm		38.4 ± 2.86 [b]
Alive *C. demersum*		52.6 ± 3.63 [c]
Control	NRR (neonates per female^{-1})	18.1 ± 0.67 [a]
Exudates		28.7 ± 1.81 [b]
POM < 25 µm		24.5 ± 1.13 [ab]
Alive *C. demersum*		41.1 ± 2.63 [c]
Control	GT (days)	20.6 ± 0.65 [ab]
Exudates		22.3 ± 0.4 [ac]
POM < 25 µm		19.2 ± 0.36 [b]
Alive *C. demersum*		24.6 ± 1.05 [c]
Control	RPI (*r*)	0.15 ± 0.06 [a]
Exudates		0.24 ± 0.02 [b]
POM < 25 µm		0.18 ± 0.07 [a]
Alive *C. demersum*		0.22 ± 0.01 [b]

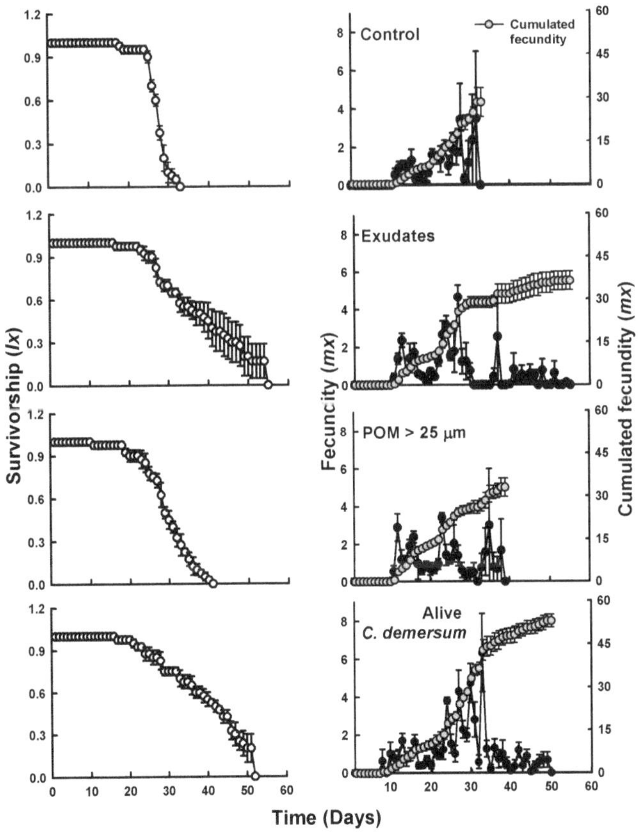

Figure 3. Survivorship (**left**) and fecundity (**right**) of *Simocephalus* cf. *mixtus* under the influence of exudates, particulate organic matter smaller than 25 µm (POM < 25 µm) + exudates, and living *Ceratophyllum demersum*. Mean ± SE data are shown based on four replicates.

4. Discussion

Population growth experiments with cladocerans have shown that submerged macrophytes indeed increased *Simocephalus* abundance as hypothesized, while the abundance of *Daphnia* was only marginally affected. Previous studies have presented varied responses of cladocerans to submerged macrophytes; however, the majority of research has focused on life table variables and the behavior of pelagic *Daphnia* [9,16–18,21], with limited attention given to *Simocephalus* and population growth responses [22,23,33].

The chemical properties of exudates derived from different macrophytes can vary significantly, leading to diverse effects on organisms [34]. For instance, *Myriophyllum spicatum* exhibits strong negative effects on cyanobacteria and green algae species through the exudation of allelopathic compounds like tellimangrandin II and ellagic acid. Similarly, *Ceratophyllum demersum* releases allelopathic sulfur compounds targeting diatoms [12,13]. In a meta-analysis examining the inhibition mechanisms of submerged macrophytes on algae, Liu et al. [35] found that the inhibition of algae by submerged plants is species-specific, with *Ceratophyllum demersum* demonstrating the strongest allelopathic effect on algae.

When assessing the impact of macrophytes on cladocerans, Gutierrez and Mayora [15] emphasized variations in phenols and chromophoric dissolved organic matter (DOM) exudates among different macrophyte species, resulting in changes in the avoidance behavior of cladocerans and copepods. Additionally, various species of zooplankton exhibited divergent responses to the same species of macrophyte. Moreover, additional research indicates that the association between macrophytes and cladocerans varies depending on the particular species involved [15,36]. Therefore, the effect of macrophyte exudates relies on both the specific macrophyte species and the susceptibility of plankton.

Most research on macrophyte–*Daphnia* interactions has revealed negative impacts of macrophytes on *Daphnia*, attributed to physical and chemical factors such as accelerated sedimentation of food particles due to macrophyte structure, macrophytes acting as barriers to swimming [18], and *Daphnia* avoiding macrophytes due to heightened predation risk from fish [15,19,21]. Additionally, numerous studies indicate that allelopathic substances exuded by macrophytes can reduce fecundity, delay maturation, and diminish *Daphnia* growth [9]. However, the effects of DOM exudated from macrophytes on cladocerans have been poorly assessed in laboratory conditions [15].

Burks et al. [17] found that *Daphnia* exposed to chemical exudates from macrophytes experienced delayed maturation and reduced fecundity, implying indirect costs to *Daphnia* due to the allelopathic inhibition of their algal food source. However, in our investigation, distinct negative impacts on the cladoceran population were not evident; instead, we observed a prolonged lag phase in the growth population of *Daphnia* exposed to *Stuckenia pectinata*; consequently, the growth rate was diminished, even though the abundances across all treatments remained similar.

On the other hand, the advantageous impacts of macrophyte exudates on *Simocephalus* spp. were evidenced by the enhancement of lifespan, survivorship, and fecundity through allelochemicals released by *Egeria densa* [22]. These findings align with our own, revealing a positive influence of submerged macrophyte exudates on *Simocephalus* demography and filtration rates. Notably, age-specific survivorship showed a 40% increase, accompanied by elevated average lifespan and reproductive rates in treatments involving exudates and alive plants. Additionally, positive effects on the population density of *Macrothrix triserialis*, *Diaphanosoma birgei*, *Simocephalus mixtus*, and *Daphnia mendotae* from *Egeria densa*'s allelochemicals were reported as well [33].

Despite these advancements, our understanding of the effects of macrophyte exudates and secondary metabolites on aquatic herbivores remains limited [37]. Some explanations attribute these outcomes to hormesis, occurring under low-stress conditions where reduced toxicity promotes reproduction, longevity, and survival [38]. These conditions are often found in littoral vegetated areas, potentially explaining the high zooplankton abundances observed there. Moreover, allelochemicals released by macrophytes have both negative and positive effects on different species of algae [39], suggesting a similar dual impact on zooplankton.

From an eco-evolutionary perspective, *Simocephalus* has coevolved in the littoral with presence of macrophytes [40], where allelopathic exudates are prevalent. Consequently, this has shaped the conditions of its habitat, leading to lower fitness in the absence of such conditions, as observed in the control group. While the survivorship and fecundity of *Simocephalus* increased in the presence of exudates and live *Ceratophyllum*, a longer generation time indicated a trade-off between these variables. The presence of phenolic compounds and tannins negatively affected herbivores' feeding rates, yet some secondary metabolites, including tannins, exhibited beneficial effects against pathogens and stressors, akin to their roles in terrestrial herbivores [37]. Additionally, phenolic compounds are recognized as antioxidant compounds [41], and recent studies have identified anticancer phenolic compounds in *C. demersum* [42], underscoring macrophytes as a rich source of beneficial substances. However, our knowledge regarding the benefits of secondary metabolites exuded from macrophytes in aquatic systems remains sparse [37]. Thus, identifying and studying these beneficial secondary metabolites in macrophyte exudates presents a critical opportunity to understand processes in freshwater ecology.

Macrophyte exudates constitute a substantial portion of total dissolved organic matter (DOM) in vegetated habitats [15], and numerous studies elucidate their role in enhancing primary and secondary productivity [43]. Mesocosm experiments conducted by Balzer et al. [44] delineate how allochthonous DOM stimulates autotrophs and mixotrophs, playing a pivotal role in zooplanktonic secondary production. The DOM generated by submerged macrophytes primarily comprises protein-like substances exhibiting high activity, with discernible variations observed among different aquatic plants. Some investigations have documented that DOM significantly influences alterations in microbial community composition [43]. In our population growth experiments, the treatments involving particulate organic matter smaller than 25 μm (POM < 25 μm) were statistically comparable to exudates concerning *Simocephalus* and *Daphnia*. This suggests that the microbiota and detritus associated with macrophytes may not serve as a substantial nutritional food source for cladocerans under these conditions.

Feeding behavior in daphnids is characterized by a generalist approach, wherein the largest ingestible particle size is determined by the frontal opening of the carapace, while the smallest particles are limited by the mesh size of the filtration setules in the thoracopods [45]. Mesh sizes vary among different cladoceran species, ranging from 0.16 μm for *Diaphanosoma* to 4.7 μm in *Sida*. For *Daphnia* species, mesh size ranges from 0.23 to 0.45 μm for *D. cucullata* up to 0.56 to 1.8 μm for *D. hyalina* [46]; however, data for *Simocephalus mixtus* are currently unavailable. Despite their ability to ingest a wide range of particle sizes, the optimal food size range for cladocerans typically spans from 3 to 20 μm [45]. Larger cladocerans such as *D. pulex* exhibit up to threefold higher filtration efficiency on 20 μm algal particles compared to bacteria, which are usually smaller than 2 μm, thereby safeguarding picoplankton from extensive grazing by large cladocerans [45,47]. Consequently, the bacteria associated with submerged macrophytes and produced through exudated dissolved organic matter (DOM) may not be efficiently consumed by the relatively large *Simocephalus* cf. *mixtus* (1781 ± 163 μm) and *Daphnia* cf. *pulex* (2371 ± 155 μm) utilized in our experiments.

Detritus represents another crucial component of POM. For littoral *Simocephalus*, the fatty acid composition aligns with that of littoral particulate matter, underscoring its relevance as a resource for littoral species. In contrast, *Daphnia* primarily derives nutrients from phytoplankton [48]. Cladocerans typically recycle DOM through the microbial loop, wherein they readily consume flagellates and certain ciliates that prey on bacteria [46]. However, our observations in the aquarium containing *C. demersum* revealed predominantly amoebas and hypotrich ciliates among the protozoans, which are less susceptible to consumption by filtering cladocerans due to their surface association [49].

In terms of filtration and consumption rates, *D.* cf. *pulex* demonstrated higher cell consumption and water filtration compared to *S.* cf. *mixtus* in control conditions, consistent with findings from previous research [50]. This discrepancy can be attributed to the larger size of *Daphnia* relative to *Simocephalus* [51]. The filtration rates of cladocerans are influenced by factors such as food quality and quantity, particle size, and the size of the cladoceran itself [48]. Therefore, we anticipated higher filtration rates in *Daphnia* compared to *Simocephalus*, as reflected in our control data. However, *Simocephalus* exposed to exudates from *Ceratophyllum* and *Stuckenia* showed a statistically significant increase in filtration rates, approaching the maximum reported values for its size [51]. In contrast, the filtration rates of *Daphnia* were not statistically affected by macrophyte exudates and remained consistent with previous reports [52].

Despite the crucial role of filter feeders and the filtration process in maintaining water clarity [53], there is limited information available regarding the effects of macrophyte exudates on cladoceran filtration and feeding rates. Typically, the filtration rates of cladocerans range from 0.9 to 135 mL per individual over a 24-h period [48], which aligns with our data for both *Daphnia* and *Simocephalus*. In summary, the presence of macrophytes and their exudates can influence the filtration and consumption rates of *Simocephalus*, but not *Daphnia*. This has significant implications for managing algal blooms and preserving water quality in aquatic ecosystems.

5. Conclusions

In this study, we investigated the effects of three submerged macrophytes, their exudates, and associated particulate organic matter on the population growth and filtration rates of *Simocephalus* and *Daphnia*, highlighting the complexity of interactions between aquatic plants and zooplankton. Our results reveal significant differences in the responses of different cladoceran species to macrophyte exudates, emphasizing the importance of considering species-specific responses in ecological studies and biomanipulation trials. Submerged macrophytes had positive effects on the abundance and rate of population increase of *Simocephalus* cf. *mixtus* through the exudation of chemical substances, with *Ceratophyllum demersum* showing the strongest positive effect on survivorship, average lifespan, and reproductive rates of *S.* cf. *mixtus*, with no clear negative effects on *Daphnia* cf. *pulex*. This differential response underscores the specificity of interactions between zooplankton and macrophyte exudates. The increase in filtration and feeding rates in the presence of macrophyte exudates may partially explain the positive effects registered on demographic parameters. The findings of this study have implications beyond individual species dynamics, encompassing ecological consequences. The influence of macrophyte exudates on cladoceran populations may have cascading effects on water quality and the control of algal proliferation. Understanding these complex interactions is crucial for effective ecosystem management and restoration efforts.

In summary, our study provides valuable insights into the ecological importance of macrophyte exudates in freshwater ecosystems. By elucidating the underlying mechanisms of zooplankton responses to macrophyte-derived compounds, we lay the groundwork for future research.

Author Contributions: Conceptualization: C.A.E.-R., A.L.-V. and M.d.R.S.-R.; formal analysis: C.A.E.-R., A.L.-V. and L.J.M.-C.; funding acquisition: A.L.-V. and M.d.R.S.-R.; methodology: C.A.E.-R., L.J.M.-C., I.M.S.-M., M.d.R.S.-R., L.P.-S. and L.R.-D.l.P.; writing—original draft: C.A.E.-R., A.L.-V., L.J.M.-C., I.M.S.-M., M.d.R.S.-R. and L.R.-D.l.P.; writing—review and editing: C.A.E.-R., A.L.-V., L.J.M.-C., I.M.S.-M., M.d.R.S.-R., L.P.-S. and L.R.-D.l.P. All authors have read and agreed to the published version of the manuscript.

Funding: This study was supported by PAPIIT IN231920 Project, UNAM.

Data Availability Statement: Data are available from the authors upon reasonable request.

Acknowledgments: We thank Biol. Christian Eduardo Torres-Sánchez and to PAPIIT Project IN231920. LJMC thanks to PCML for master studies and CONAHCyT (CVU 1313323) for a scholarship.

Conflicts of Interest: The authors declare no conflicts of interest.

References

1. Moss, B. *Ecology of Freshwaters: A View for the Twenty-First Century*; Wiley-Blackwell: Oxford, UK, 2010; p. 480.
2. Downing, A.J. Limnology and oceanography: Two estranged twins reuniting by global change. *Inland Waters* **2014**, *4*, 215–232. [CrossRef]
3. Gulati, R.D.; Pires, M.D.; van Donk, E. Restoration of freshwater lakes. In *Restoration Ecology: The New Frontier*, 2nd ed.; van Andel, J., Aronson, J., Eds.; Wiley-Blackwell: Oxford, UK, 2012; pp. 233–247. [CrossRef]
4. Janssen, A.B.G.; Hilt, S.; Kosten, S.; de Klein, J.J.M.; Paerl, H.W.; van de Waal, D.B. Shifting states, shifting services: Linking regime shifts to changes in ecosystem services in shallow lakes. *Freshw. Biol.* **2021**, *66*, 1–12. [CrossRef]
5. Sondergaard, M.; Jeppesen, E.; Lauridsen, T.L.; Skov, C.; van Nes, E.H.; Roijackers, R.M.M.; Lammens, E.; Portielje, R. Lake restoration: Successes, failures, and long-term effects. *J. Appl. Ecol.* **2007**, *44*, 1095–1105. [CrossRef]
6. Xu, S.; Lyu, P.; Zheng, X.; Yang, H.; Xia, B.; Li, H.; Zhang, H.; Ma, S. Monitoring and control methods of harmful algal blooms in Chinese freshwater system: A review. *Environ. Sci. Pollut. Res.* **2022**, *29*, 56908–56927. [CrossRef] [PubMed]
7. Klapper, H. Technologies for Lake Restoration. *J. Limnol.* **2003**, *62*, 73–90. [CrossRef]
8. Dodds, W.; Whiles, M. *Freshwater Ecology: Concepts and Environmental Applications of Limnology*, 2nd ed.; Elsevier: Amsterdam, The Netherlands, 2010; p. 821.
9. Burks, R.L.; Mulderij, G.; Gross, E.; Jones, I.; Jacobsen, L.; Jeppesen, E.; van Donk, E. Center stage: The crucial role of macrophytes in regulating trophic interactions in shallow lakes. In *Wetlands: Functioning, Biodiversity Conservation, and Restoration*; Bobbink, R., Beltman, B., Verhoeven, J.T.A., Whigham, D.F., Eds.; Springer: Berlin/Heidelberg, Germany, 2006; Volume 191, pp. 37–59. [CrossRef]
10. O'Hare, W.T.; Aguiar, F.C.; Asaeda, T.; Bakker, E.S.; Chambers, P.A.; Clayton, J.S.; Elger, A.; Ferreira, T.M.; Gross, E.M.; Gunn, I.D.M.; et al. Plants in aquatic ecosystems: Current trends and future directions. *Hydrobiologia* **2018**, *812*, 1–11. [CrossRef]
11. Timms, R.M.; Moss, B. Prevention of growth of potentially dense phytoplankton populations by zooplankton grazing in the presence of zooplanktivorous fish, in a shallow wetland ecosystem. *Limnol. Oceanogr.* **1984**, *29*, 472–486. [CrossRef]
12. Hu, H.; Hong, Y. Algal-bloom control by allelopathy of aquatic macrophytes—A review. *Front. Environ. Sci. Eng. China* **2008**, *2*, 421–438. [CrossRef]
13. Erhard, D. Alleopathy in aquatic environments. In *Allelopathy*; Reigosa, M., Pedrol, N., González, L., Eds.; Springer: Dordrecht, The Netherlands, 2006; pp. 433–450.
14. Alvarez, M.F.; Benítez, H.H.; Solari, L.C.; Villegas Cortés, J.C.; Gabellone, N.A.; Claps, M.C. Effects of polyphenols on plankton assemblages and bacterial abundance representative of a pampean shallow lake: An experimental study. *Aquat. Microb. Ecol.* **2020**, *85*, 85–100. [CrossRef]
15. Gutierrez, M.F.; Mayora, G. Influence of macrophyte integrity on zooplankton habitat preference, emphasizing the released phenolic compounds and chromophoric dissolved organic matter. *Aquat. Ecol.* **2016**, *50*, 137–151. [CrossRef]
16. Burns, C.W.; Dodds, A. Food limitation, predation and allelopathy in a population of Daphnia carinata. *Hydrobiologia* **1999**, *400*, 41–53. [CrossRef]
17. Burks, R.L.; Jeppesen, E.; Lodge, D.M. Macrophyte and fish chemicals suppress Daphnia growth and alter life-history traits. *Oikos* **2000**, *88*, 139–147. [CrossRef]
18. Cerbin, S.; van Donk, E.; Gulati, R.D. The influence of Myriophyllum verticillatum and artificial plants on some life history parameters of Daphnia magna. *Aquat. Ecol.* **2007**, *41*, 263–271. [CrossRef]
19. Gutierrez, M.F.; Paggi, J.C. Chemical repellency and adverse effects of free-floating macrophytes on the cladoceran Ceriodaphnia dubia under two temperature levels. *Limnology* **2014**, *15*, 37–45. [CrossRef]
20. Song, Y.; Liew, J.H.; Sim, D.Z.H.; Mowe, A.D.; Mitrovic, S.M.; Tan, H.T.W.; Yeo, D.C.J. Effects of macrophytes on lake-water quality across latitudes: A meta-analysis. *Oikos* **2019**, *128*, 468–481. [CrossRef]
21. Meerhoff, M.; Fosalba, C.; Bruzzone, C.; Mazzeo, N.; Noordoven, W.; Jeppesen, E. An experimental study of habitat choice by Daphnia: Plants signal danger more than refuge in subtropical lakes. *Freshwater Biol.* **2006**, *51*, 1320–1330. [CrossRef]
22. Espinosa-Rodríguez, C.A.; Rivera-De la Parra, L.; Martínez-Téllez, A.; Gómez-Cabral, G.C.; Sarma, S.S.S.; Nandini, S. Allelopathic interactions between the macrophyte Egeria densa and the plankton (alga, Scenedesmus acutus and cladocerans, Simocephalus spp.): A laboratory study. *J. Limnol.* **2016**, *75*, 151–160. [CrossRef]
23. Montiel, M.A. Interacción Lirio Acuático-Plancton: Efectos Directos en el Zooplancton del Lago de Xochimilco. Ph.D. Thesis, Universidad Nacional Autónoma de México, Mexico City, Mexico, 2019.
24. Reitsema, R.E.; Meire, P.; Schoelynck, J. The future of freshwater macrophytes in a changing world: Dissolved organic carbon quantity and quality and its interactions with macrophytes. *Front. Plant. Sci.* **2018**, *9*, 629. [CrossRef]
25. Kissman, C.E.H.; Williamson, C.E.; Rose, K.C.; Saros, J.E. Nutrients associated with terrestrial dissolved organic matter drive changes in zooplankton: Phytoplankton biomass ratios in an alpine lake. *Freshwater. Biol.* **2016**, *62*, 40–51. [CrossRef]
26. Cuassolo, F.; Bastidas-Navarro, M.; Balseiro, E.; Modenutti, B. Effect of light on particulate and dissolved organic matter production of native and exotic macrophyte species in Patagonia. *Hydrobiologia* **2016**, *766*, 29–42. [CrossRef]

27. Cuevas Madrid, H.; Lugo Vázquez, A.; Peralta Soriano, L.; Morlán Mejía, J.; Vilaclara Fatjó, G.; Sánchez Rodríguez, M.d.R.; Escobar Oliva, M.A.; Carmona Jiménez, J. Identification of key factors affecting the trophic state of four tropical small water bodies. *Water* **2020**, *12*, 1454. [CrossRef]
28. Andersen, R.A.; Berjes, J.A.; Harrison, P.J.; Watanabe, M.M. Recipes for freshwater and seawater media. In *Algal Culturing Techniques*; Andersen, R.A., Ed.; Elsevier Academic Press: London, UK, 2005; pp. 429–438.
29. Weber, C.I. *Methods for Measuring the Acute Toxicity of Effluents and Receiving Waters to Freshwater and Marine Organisms*, 4th ed.; United States Environmental Protection Agency: Cincinnati, OH, USA, 1993.
30. Krebs, C. *Ecology: The Experimental Analysis of Distribution and Abundance*, 6th ed.; Ashford Color Press: Harlow, UK, 2014; p. 646.
31. Sokal, R.R.; Rohlf, F.L. *Biometry: The Principles and Practice of Statistics in Biological Research*, 3rd ed.; W.H. Freeman and Company: San Francisco, CA, USA, 1995; p. 887.
32. Rigler, F.H. Feeding rates: Zooplankton. In *A Manual for the Assessment of Secondary Productivity in Freshwaters*; Edmondson, W.T., Winberg, G.G., Eds.; Blackwell Scientific: Oxford, UK, 1971; pp. 228–255.
33. Espinosa-Rodríguez, C.A.; Sarma, S.S.S.; Nandini, S. Effect of the allelochemicals from the macrophyte Egeria densa on the competitive interactions of pelagic and littoral cladocerans. *Chem. Ecol.* **2017**, *33*, 247–256. [CrossRef]
34. Wilkins, K.W.; Overholt, E.; Williamson, C. The effects of dissolved organic matter from a native and an invasive plant species on juvenile Daphnia survival and growth. *J. Plankton Res.* **2020**, *42*, 453–456. [CrossRef]
35. Liu, X.; Sun, T.; Yang, W.; Li, X.; Ding, J.; Fu, X. Meta-analysis to identify inhibition mechanisms for the effects of submerged plants on algae. *J. Environ. Manag.* **2024**, *355*, 120480. [CrossRef] [PubMed]
36. Kurbatova, S.A.; Lapteva, N.A.; Bykova, S.N.; Yershov, I.Y. Aquatic plants as a factor that changes trophic relations and the structure of zooplankton and microperiphyton communities. *Biol. Bull.* **2019**, *46*, 284–293. [CrossRef]
37. Gross, E.M.; Bakker, E.S. The role of plant secondary metabolites in freshwater macrophyte–herbivore interactions: Limited or unexplored chemical defenses? In *The Ecology of Plant Secondary Metabolites: From Genes to Global Processes*; Iason, G.R., Dicke, M., Hartley, S.E., Eds.; Ecological Reviews; Cambridge University Press: Cambridge, UK, 2012; pp. 154–169.
38. Calabrase, E.J.; Baldwin, L.A. Hormesis as a biological hypothesis. *Environ. Health Persp.* **1998**, *106*, 357–362. [CrossRef]
39. Harris, T.D.; Reinl, K.L.; Azarderakhsh, M.; Berger, S.A.; Castro Berman, M.; Bizic, M.; Bhattacharya, R.; Burnet, S.H.; Cianci-Gaskill, J.A.; de Senerpont Domis, L.N.; et al. What makes a cyanobacterial Bloom disappear? A review of the abiotic and biotic cyanobacterial bloom factors. *Harmful Algae* **2024**, *133*, 102599, *in press*. [CrossRef] [PubMed]
40. Orlova-Bienkowskaja, M. *Guides to the Identification of the Microinvertebrates of the Continental Waters of the World. Cladocera: Anomopoda. Daphniidae: Genus Simocephalus*; Backhuys Publishers: Leiden, The Netherlands, 2001; p. 130.
41. Susanti, I.; Pratiwi, R.; Rosandi, Y.; Hasanah, A.N. Separation methods of phenolic compounds from plant extract as antioxidant agents candidate. *Plants* **2024**, *13*, 965. [CrossRef] [PubMed]
42. Maslyk, M.; Lenard, T.; Olech, M.; Martyna, A.; Poniewozik, M.; Boguszewska-Czubara, A.; Kochanowicz, E.; Czubak, P.; Kubiński, K. Ceratophyllum demersum the submerged macrophyte from the mining subsidence reservoir Nadrybie Poland as a source of anticancer agents. *Sci. Rep.* **2024**, *14*, 6661. [CrossRef]
43. Ren, H.; Wang, G.; Ding, W.; Li, H.; Shen, X.; Shen, D.; Jiang, X.; Qadeer, A. Response of dissolved organic matter (DOM) and microbial community to submerged macrophytes restoration in lakes: A review. *Environ. Res.* **2023**, *231*, 116185. [CrossRef]
44. Balzer, M.J.; Hitchcock, J.N.; Hadwen, W.L.; Kobayashi, T.; Westhorpe, D.P.; Boys, C.; Mitrovic, S.M. Experimental additions of allochthonous dissolved organic matter reveal multiple trophic pathways to stimulate planktonic food webs. *Freshw. Biol.* **2023**, *68*, 821–836. [CrossRef]
45. Lampert, W.; Sommer, U. *Limnoecology: The Ecology of Lakes and Streams*, 2nd ed.; Oxford University Press: New York, NY, USA, 2007; p. 324.
46. Geller, W.; Müller, H. The filtration apparatus of Cladocera: Filter mesh-sizes and their implications on food selectivity. *Oecologia* **1981**, *49*, 316–321. [CrossRef]
47. Knoechel, Y.; Holtby, B. Cladoceran filtering rate: Body length relationships for bacterial and large algal particles. *Limnol. Oceanogr.* **1986**, *31*, 195–199. [CrossRef]
48. Smirnov, N.N. *Physiology of the Cladocera*, 2nd ed.; Academic Press: Cambridge, MA, USA, 2017; p. 418.
49. Fenchel, T. *The ecology of Protozoa*; Springer: Berlin/Heidelberg, Germany, 1987; p. 197.
50. Navarro, L.; Rejas, D. Potential of four native zooplankton species for biomanipulation of eutrophic lakes in the Valley of Cochabamba, Bolivia. *Rev. Biol. Ecol. Cons. Amb.* **2009**, *25*, 1–9.
51. Burns, C.W. Relation between filtering rate, temperature, and body size in four species of Daphnia. *Limnol. Oceanogr.* **1969**, *14*, 693–700. [CrossRef]
52. Serra, T.; Müller, M.F.; Barcelona, A.; Salvadó, V.; Pous, N.; Colomer, F. Optimal light conditions for Daphnia filtration. *Sci. Total Environ.* **2019**, *686*, 151–157. [CrossRef]
53. Scheffer, M. The effect of aquatic vegetation on turbidity; how important are the filter feeders? *Hydrobiologia* **1999**, *408*, 307–316. [CrossRef]

plants

MDPI

Article

Resistance to *Frankliniella occidentalis* during Different Plant Life Stages and under Different Environmental Conditions in the Ornamental Gladiolus

Dinar S. C. Wahyuni [1,2], Peter G. L. Klinkhamer [1], Young Hae Choi [3,4] and Kirsten A. Leiss [5,*]

1 Plant Science and Natural Products, Institute of Biology (IBL), Leiden University, Sylviusweg 72, 2333BE Leiden, The Netherlands; dinarsari_cw@staff.uns.ac.id (D.S.C.W.); p.g.l.klinkhamer@biology.leidenuniv.nl (P.G.L.K.)
2 Pharmacy Department, Faculty Mathematics and Natural Sciences, Universitas Sebelas Maret, Jl. Ir. Sutami 36A, Surakarta 57126, Indonesia
3 Natural Products Laboratory, Institute of Biology, Leiden University, Sylviusweg 72, 2333BE Leiden, The Netherlands; y.choi@chem.leidenuniv.nl
4 College of Pharmacy, Kyung Hee University, Seoul 02447, Republic of Korea
5 Business Unit Horticulture, Wageningen University and Research Center, Postbus 20, 2665ZG Bleiswijk, The Netherlands
* Correspondence: kirsten.leiss@wur.nl; Tel.: +31-3174-81747

Abstract: The defense mechanisms of plants evolve as they develop. Previous research has identified chemical defenses against Western flower thrips (WFT) in *Gladiolus* (*Gladiolus hybridus* L.). Consequently, our study aimed to explore the consistency of these defense variations against WFT across the various developmental stages of *Gladiolus* grown under different conditions. Thrips bioassays were conducted on whole plants at three developmental stages, using the Charming Beauty and Robinetta varieties as examples of susceptible and resistant varieties, respectively. Metabolomic profiles of the leaves, buds and flowers before thrips infestation were analyzed. The thrips damage in Charming Beauty was more than 500-fold higher than the damage in Robinetta at all plant development stages. Relative concentrations of triterpenoid saponins and amino acids that were associated with resistance were higher in Robinetta at all plant stages. In Charming Beauty, the leaves exhibited greater damage compared to buds and flowers. The relative concentrations of alanine, valine and threonine were higher in buds and flowers than in leaves. The Metabolomic profiles of the leaves did not change significantly during plant development. In addition, we cultivated plants under different environmental conditions, ensuring consistency in the performance of the two varieties across different growing conditions. In conclusion, the chemical thrips resistance markers, based on the analysis of vegetative plants grown in climate rooms, were consistent over the plant's lifetime and for plants grown under field conditions.

Keywords: plant development stages; *Frankliniella occidentalis*; climate- and field-grown plants; Gladiolus; eco-metabolomics

1. Introduction

Plant defenses are not fixed throughout a plant's life. Major changes occur depending on growing conditions, plant development and the level of biotic and abiotic stress. For breeders, such changes may present a problem when they want to detect robust chemical markers for resistance in their breeding programs.

Plant resistance to herbivores has mostly been studied under controlled conditions in growth cabinets or climate rooms to minimize the effects of external variables on the plant metabolome. Under laboratory conditions, photoperiod, light intensity, temperature and humidity are controlled, whereas in the field, those conditions are highly variable. These external variables may thus cause variation in the levels of defense compounds and consequently affect plants' resistance to herbivores. For instance, the concentration of triterpenoid saponins in plants is affected by habitat, season, plant age, light, temperature and water [1]. Amino acid levels were reported to depend on light conditions [2]. Also, drought affects amino acid contents and, through this, herbivore feeding performance [3].

During a plant's lifetime, major changes in its defense system occur. This can be the result of aging tissues [4], or these changes can be associated with developmental switches such as from seedling to vegetative stage or from vegetative to flowering stage [5]. Generally, it is assumed that plant parts that most strongly contribute to s fitness are defended best [6]. For instance, young leaves are, in general, better protected from generalist herbivores than older leaves [7], and buds and flowers are better protected than leaves [8]. The ultimate choice of herbivores will be determined by both the nutritional value of the tissue and its level of defense. While for herbivores, such as thrips, young flowers with pollen can have high nutritional value [9], they may at the same time be better protected and have accumulated higher defense levels than other plant tissues such as leaves [9]. The effect of developmental stage or plant age on resistance has been well studied for a number of insect herbivores, including Western flower thrips (*Frankliniella occidentalis*, WFT). The WFT preference pattern is not fully consistent across plant species. In a greenhouse study with *Impatiens walleriana*, the rank order of WFT preference was 1. plants with flower buds, 2. plants with fully opened flowers with pollen, 3. plants with fully opened flowers without pollen and 4. plants with foliage without flowers [10]. In *Calystegia sepium*, WFT numbers increased during bud development and opening and reached a peak just before flowers started to wilt [11]. In both *Impatiens wallerana* and *Calystegia sepium*, WFT preferred flowers over leaves [10,11]. In tomato [12] and in Senecio [13], WFT damage was greater in older leaves. In tomato, this difference became stronger after external application of JA. Although from an evolutionary point of view, it makes sense that tissues that contribute less to fitness are not optimally defended, this presents a problem to growers. While high infestation levels on older leaves may not reduce flower or seed production, they may lead to unmarketable products or higher levels of virus infections as, e.g., in the case of thrips [14].

For plant breeders, potential changes in the plant's defense system during plant development presents a problem because selection in breeding programs is based on the analyses of early life stages. The question is whether or not predictions of resistance in young plants are good indicators of resistance later in life. Especially, for herbivores that show a clear preference for particular plant organs such as buds, flowers or seeds this question is highly relevant. In this paper we studied the defenses of *Gladiolus* against WFT at three developmental stages and under different growing conditions. WFT is one of the most serious pests of agricultural and horticultural crops worldwide [15], causing losses of millions of euros. WFT is highly polyphagous, invading fruit, vegetables and ornamentals [16]. Thrips have piercing–sucking mouthparts which allow them to feed on different types of plant cells [17]. After sucking up the cell's content, these fill with air, leading to the characteristic silver damage. Moreover, they are vectors of viral diseases [14].

In *Gladiolus* too, thrips infestation presents a severe problem. Plant breeders are in need of morphological or chemical markers to assist their breeding programs and to make full use of the natural variation that is present in *Gladiolus* with respect to thrips resistance. In earlier work, we investigated the differences in WFT resistance of various *Gladiolus* varieties [18] grown under climate room conditions. We detected, in a multivariate analysis of NMR data, signals related to thrips resistance. These were a signal at δ 0.90 ppm linked to triterpenoid saponins and the amino acids alanine and threonine. Subsequent correlation analyses gave significant relationships with the signal of 0.90 ppm, linked to triterpenoid saponins, alanine and threonine. All these signals were highly correlated among each other and with density of papillae [18]. Most likely these defence compounds are produced and/or stored in the extra cuticular papillae.

The objective of this study was to investigate the effect of plant developmental stages and environmental conditions on plant resistance to WFT. For this we investigated *Gladiolus* plants grown under natural field conditions of a plant breeder, plants transferred from the field (to a climate room and plants grown during the whole experiment in a climate room). The vegetative life stage comprises about 80% of the total life cycle of Gladiolus. However, success in later developmental stages of the plants is crucial for bulb and flower production. We, therefore, compared metabolomic profiles and WFT infestation for three developmental stages: vegetative, generative stage with buds and generative stage with flowers.

For our experiments we used the *Gladiolus* varieties Robinetta and Charming Beauty which in the vegetative stage were shown to be highly resistant and susceptible to WFT, respectively [18]. We specifically addressed the following questions:

- Do Robinetta and Charming Beauty show consistent differences in WFT resistance over all development stages?
- Does WFT damage differ between plant organs?
- Does WFT damage to leaves differ among plant development stages?
- Are differences between the metabolomic profiles of Robinetta and Charming Beauty consistent across developmental stages?
- Do the concentrations of defence compounds related to WFT resistance differ among plant organs?
- Do the concentrations of compounds that were related to WFT resistance alter with the development stages of the plant?
- Are the metabolic profiles of the plants dependent on the environmental conditions?
- And if so: Is there a change in the concentration of compounds related to thrips resistance?

2. Results

2.1. Thrips Whole Plant Bioassay

Differences in total WFT damage between the two varieties. While Charming Beauty showed considerable damage on flowers and leaves (Figure 1), hardly any damage occurred in Robinetta at all developmental stages. Total WFT damage differed significantly between Charming Beauty and Robinetta (U = 55.000, df = 1, $p = 0.000$) (Figure 2). The average total damage across all three developmental stages was: 565.22 ± 77.4 mm^2 in Charming Beauty and 3.3 ± 1.9 mm^2 in Robinetta.

(A) (B)

Figure 1. Plant silver damage in Charming Beauty on flowers (**A**) and leaves (**B**).

WFT Damage in Different Plant Organs. In Charming Beauty damage to buds accounted for 35% of the total damage in the bud stage. Damage to flowers accounted for 16% from the total damage in this stage, while no damage to buds occurred in this stage. In Robinetta damage to all plant organs was low and in buds and flowers it was even zero (Figure 2).

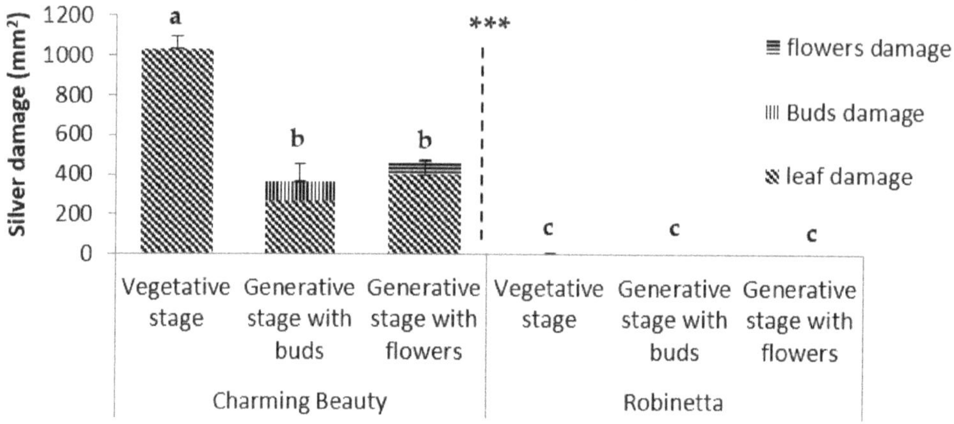

Figure 2. Plant silver damage (mm^2) in Charming Beauty and Robinetta at three plant development stages: vegetative, generative with buds and generative with flowers as measured by a whole plant Western Flower Thrips non-choice bioassay. Bars represent total plant damage, patterns within bars represent different plant organs. Differences in total plant damage within the three developmental stages were tested with one-way ANOVA (Charming Beauty) and Kruskal-Wallis (Robinetta). Data represent mean and standard errors for three to four replicates. Different letters above the bars refer to significant differences within development stages at the 0.05 level. *** Indicate significant differences between the varieties ($p < 0.000$).

WFT Damage on Leaves at Different Plant-Stages. WFT damage on leaves differed significantly among the three plant development stages in Charming Beauty (F = 16.593, df = 2, $p = 0.023$) (Figure 2). Damage in the vegetative stage was two times higher than in the generative stage with buds or flowers. In Robinetta WFT damage at all three developmental stages were close to zero and did not differ significantly developmental stages (H = 2.333, df = 2, $p = 0.311$) (Figure 2).

2.2. Metabolomic Profiling

Differences in Metabolite Profiles of Leaves Between Varieties. PCA is an unsupervised method which enables to identify the differences or similarities among samples. Charming Beauty and Robinetta differed in their leaf metabolomic profiles at all plant stages although the differences in flowers were relatively small (Figure 3A). The separation was mainly due to PC1 which explained 41% of the variation in leaf metabolites. The loading plot showed that the signals in the region between δ 1.92–0.80 ppm had a low score and thus were associated with Robinetta the WFT resistant variety (Figure 3B). The signals at δ 1.28 (signal A) and 0.90 ppm (signal B) were related to triterpenoid saponins. In this region we could further identify signals related to the amino acids valine (δ 1.06) alanine (δ 1.48), and threonine (δ 1.32). Signals with a high score on the loading plot, that thus were associated with Charming Beauty, were in the sugar region δ 5.0–3.0 ppm (Figure 3B). However, the relative concentrations of the sugars we could identify, sucrose (δ 5.40), α-glucose (δ 5.20) and β-glucose (δ 4.60) did not differ significantly between Charming Beauty and Robinetta (F = 1.284, df = 1, $p = 0.272$; F = 0.351, df = 1, $p = 0.561$ and F = 0.219, df = 1, $p = 0.645$, respectively) (Figure 4). The relative concentrations of the triterpenoid saponins that were related to signal A and signal B were significantly higher in Robinetta (H = 16.323, df = 1, $p = 0.000$ and H = 14.449, df = 1, $p = 0.000$, respectively) than in Charming Beauty (Figure 4). The relative concentrations of alanine, valine and threonine were about three to four times higher in Robinetta than in Charming Beauty (F = 73.702, df = 1, $p = 0.000$; F = 334.108, df = 1, $p = 0.000$; F = 584.607, df = 1, $p = 0.000$, respectively) (Figure 4).

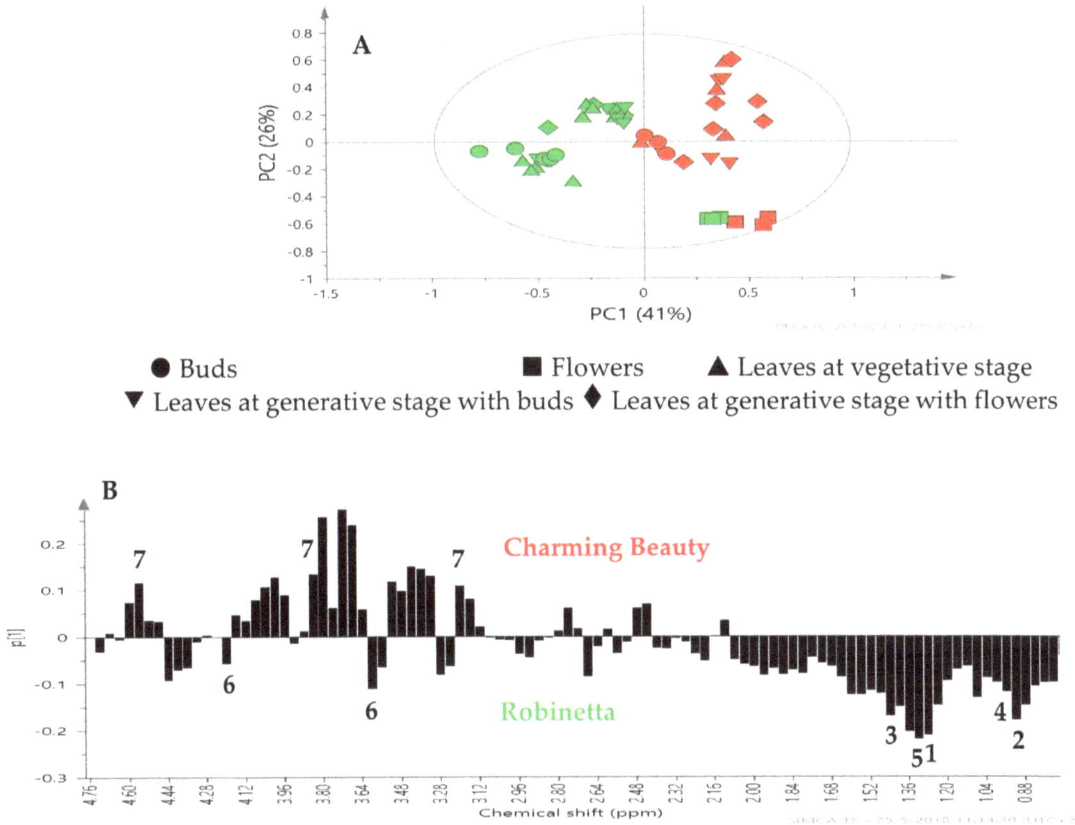

Figure 3. PCA score plot (**A**) and loading plot (**B**) for two varieties, Robinetta (green) and Charming Beauty (red) based on ^{1}H NMR spectra with (•) buds, (■) flowers (▲) leaves at vegetative stage, (▼) leaves at generative stage with buds (♦) leaves at generative stage with flowers. Metabolites are labeled as triterpenoids saponins (1 and 2), alanine (3), valine (4), threonine (5), sucrose (6) and glucose (7).

Differences in Metabolite Profiles between Plant Organs. The PCA analysis of the metabolomic profiles of the three plant organs showed clear differences for Charming Beauty (Figure 5A). PC1, which explained 42% of the variation, separated the flowers from the leaves and buds. The loading plot for PC1 showed that the region between δ 5.40–3.00 ppm which represents sugar compounds was responsible for this separation (Figure 5B). In Robinetta too plant organs were separated by their metabolomics profiles in the PCA (Figure 6A). The separation was mainly due to PC1 which explained 57% of the variation in plant metabolites. Signals with low values on the loading plot, and thus associated with buds and leaves belonged to the region δ 2.5–0.80 ppm. These signals were related to amino acids and saponins. Other signals with a negative value on the loading plot in the region δ 4.20–3.20 ppm, which we identified as being from sucrose, were associated with buds and leaves. Signals with positive values on the loading plot and thus associated with flowers, in the range from δ 4.00–3.28 ppm (Figure 6B) were identified as glucose.

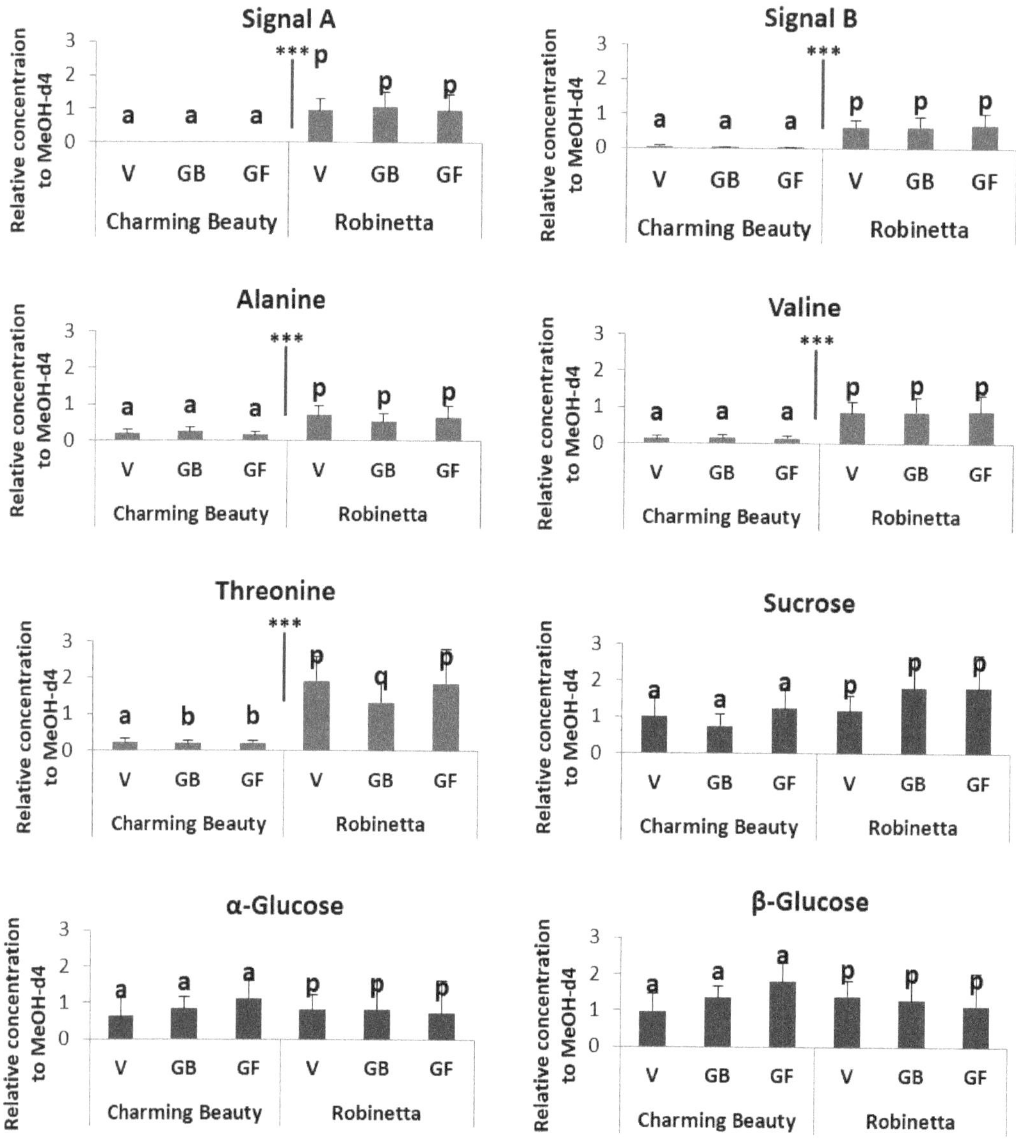

Figure 4. Relative concentration, as proportion of the internal standard, in ^{1}H NMR spectra of triterpenoid saponins (signal A and signal B), alanine, valine, threonine, sucrose, α-glucose and β-glucose in leaves of three plant development stages (Vegetative (V), Generative with buds (GB), Generative with flowers (GF)) of Charming Beauty and Robinetta. Data present the mean ± SE of four to six for replicates of leaves at the vegetative, generative with buds and generative with flower stages. Differences in relative concentrations of triterpenoid saponins and amino acids within variety and between the two varieties were analyzed by a Kruskal-Wallis test and a one-way ANOVA, respectively. Differences in the relative concentrations of sucrose, α-glucose and β-glucose between the two varieties and within variety were analyzed by one-way ANOVA. Different letters refer to significant differences among development stages within varieties at the 0.05 level. *** indicate significant differences between varieties ($p < 0.000$).

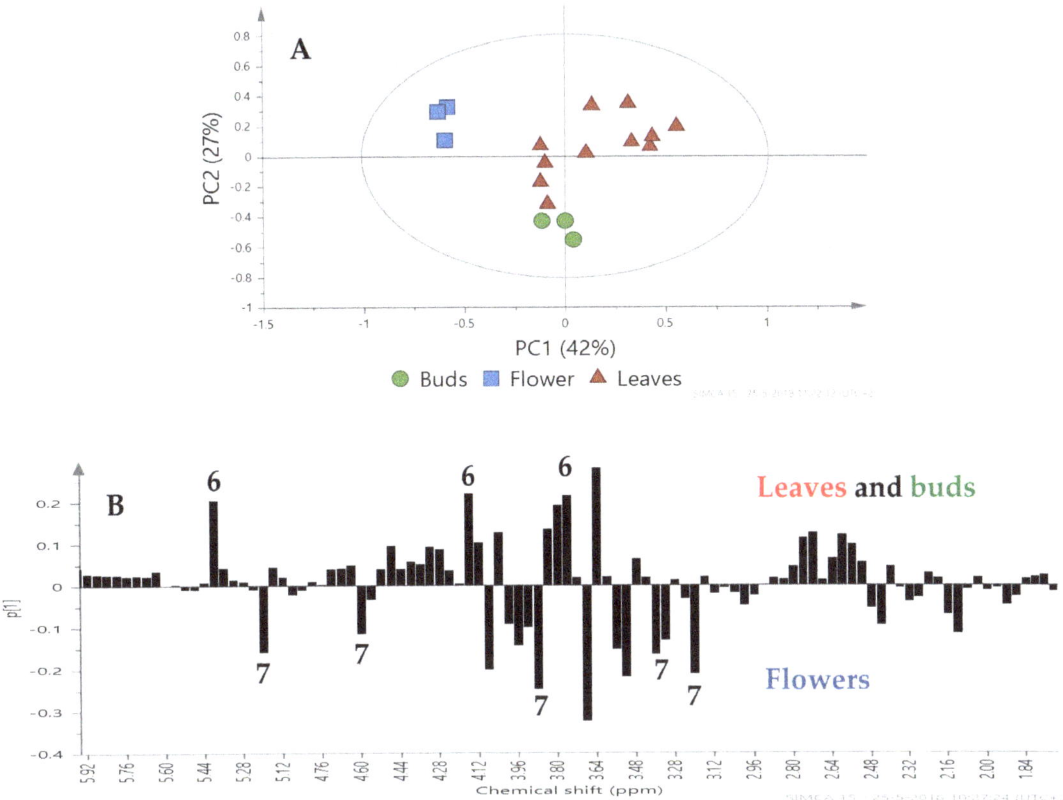

Figure 5. PCA score plot (**A**) and loading plot PC1 (**B**) for Charming Beauty based on ^{1}H NMR spectra with (•) buds (■) flowers and (▲) leaves from plants at the generative stage. Metabolites are labeled as sucrose (6) and glucose (7).

The relative concentration of signal A did not show significant differences among plant organs in Charming Beauty (H = 2.333, df = 2, p = 0.311). The relative concentration of signal B was slightly higher in buds compared to flowers and leaves (H = 6.706, df = 2, p = 0.035). Threonine, alanine and valine were two times higher in buds and flowers in Charming Beauty than in leaves (F = 5.335, df = 2, p = 0.039; F = 29.535, df = 2, p = 0.000; F = 16.347, df = 2, p = 0.002, respectively) (Figure 7). The concentrations of α- and β-glucose were about two times higher in flowers than in leaves and buds (F = 31.846, df = 2, p = 0.000 and F = 27.131, df = 2, p = 0.001, respectively) (Figure 7). However, the relative concentration of sucrose (δ 5.40 ppm) was lower in flowers than in leaves and buds (F = 5.502, df = 2, p = 0.020) (Figure 7).

In Robinetta signals A and signal B were about 50% higher in leaves and buds than in flowers (F = 63.507, df = 2, p = 0.000 and F = 14.969, df = 2, p = 0.005, respectively). Threonine was higher in leaves and buds than flowers (F = 61.767, df = 2, p = 0.000). Alanine was similar in concentration in all plant organs (F = 3.056, df = 2, p = 0.122). Valine concentration was about 50% higher in leaves and buds (F = 7.368, df = 2, p = 0.004) than in flowers (Figure 7). Relative concentrations of α- and β-glucose were about two times higher in flowers than in leaves and buds (F = 5.543, df = 2, p = 0.043 and F = 404.909, df = 2, p = 0.000, respectively) (Figure 7). In contrast, sucrose was lower in flowers than in leaves and buds (F = 10.648, df = 2, p = 0.011) (Figure 7).

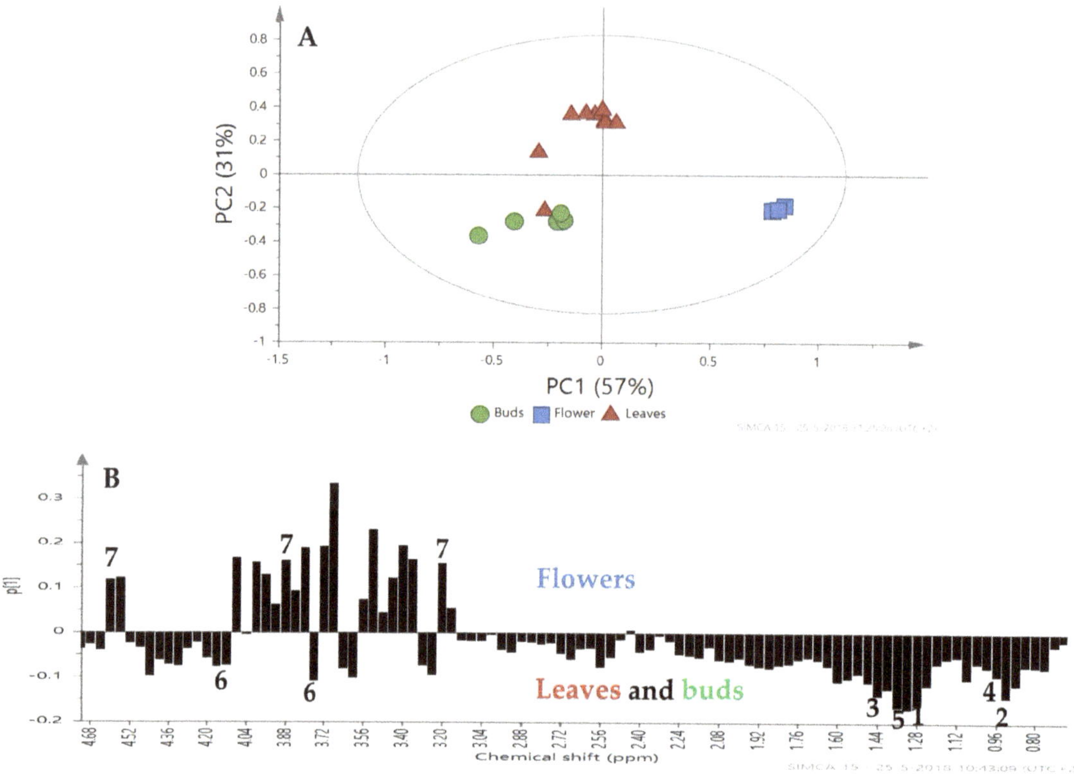

Figure 6. PCA score plot (**A**) and loading plot (**B**) for Robinetta based on ^{1}H NMR spectra with (•) buds, (■) flowers and (▲) leaves from plants in the generative stage. Metabolites are labeled as signal A (1), signal B (2), alanine (3), valine (4), threonine (5), sucrose (6) and glucose (7).

Differences in Metabolite Profiles of Leaves between Plant Development Stages. The PCA analysis of metabolomic profiles did not separate the leaves of the three developmental stages in both Charming Beauty and Robinetta. In addition, the relative concentrations of the two triterpenoid saponins (signal A and signal B), the amino acids and the sugars did not differ among the leaves from different plant developmental stages.

Differences in Metabolite Profiling between Environmental Conditions. Visual inspection of the NMR-metabolomic profiles of plants grown under different environmental conditions (field, field transition and climate room) clearly showed differences between varieties and among growing conditions (Figure 8). To further analyze these results multivariate data analysis was applied. First principal component analysis (PCA) was used. However, there was no clear clustering of the different samples within each variety. Apparently, the variability of the samples was too high to give a clear separation. Using the three growing conditions we then applied PLS-DA, for each variety. The climate room grown samples clearly separated from the other two groups of field grown plants and plants transferred from the field to the climate room. The latter two overlapped in the PLS-DA scoring plots of both Charming Beauty (Figure 9A) and Robinetta (Figure 9B). The first component explained 80% and 81% of the variance in the dataset in Charming Beauty and in Robinetta, respectively. The climate chamber-grown plants were clustered at the negative side of PC1.

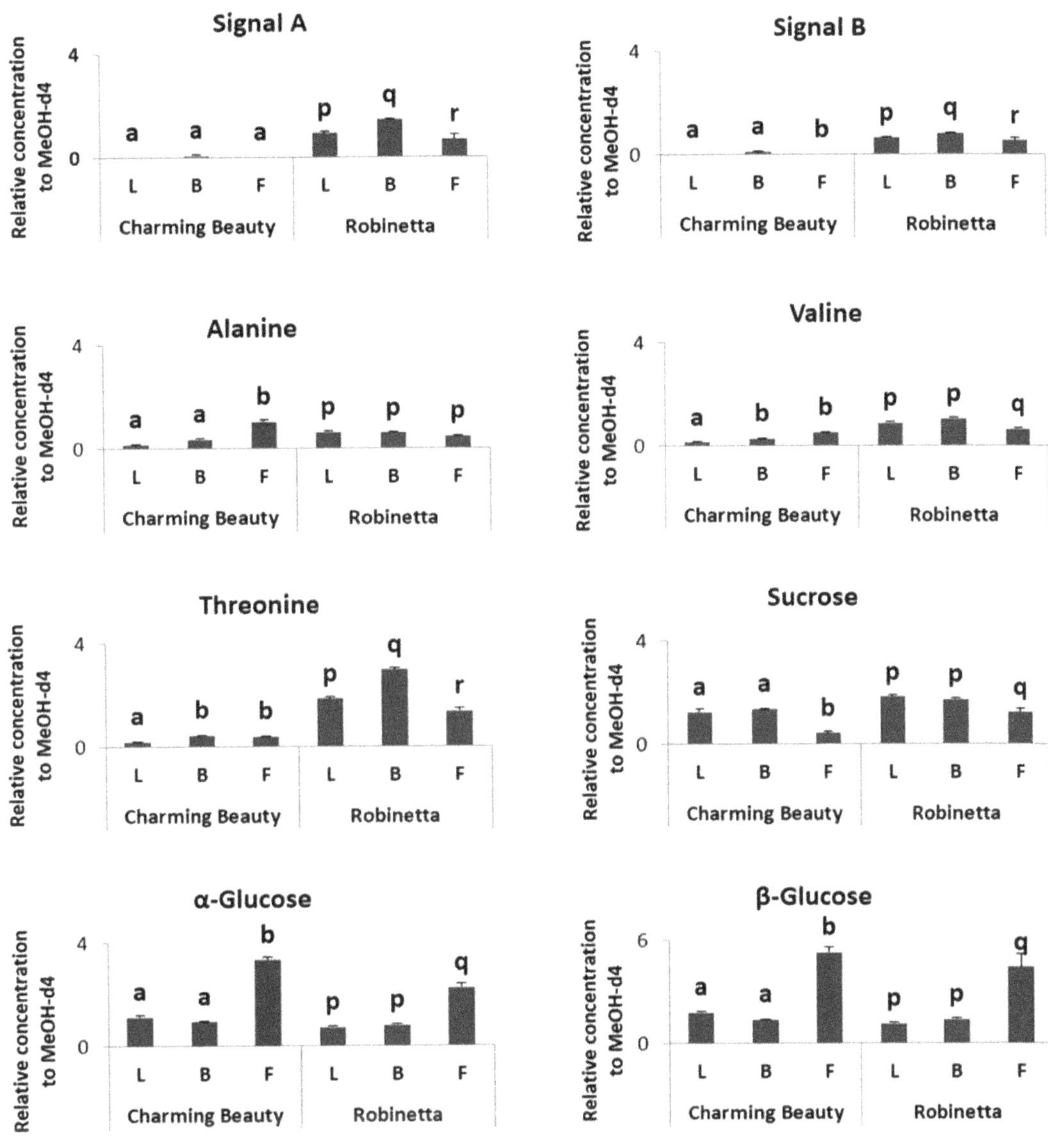

Figure 7. Relative concentrations, as proportions of the internal standard, in ^{1}H NMR spectra of triterpenoids, threonine, valine, alanine, α-glucose and β-glucose in different plant organs (Leaves (L), Buds (B), Flowers (F)) of Charming Beauty and Robinetta. Data present the mean of four to six for replicates of plants in the generative stage ± SE of the mean. Differences in relative concentrations of triterpenoid saponins and amino acids within variety and between the two varieties were analyzed by Kruskal-Wallis test and one-way ANOVA, respectively. Different letters refer to significant differences among plant organs within varieties at the 0.05 level. Differences between varieties were not significant at the 0.05 level.

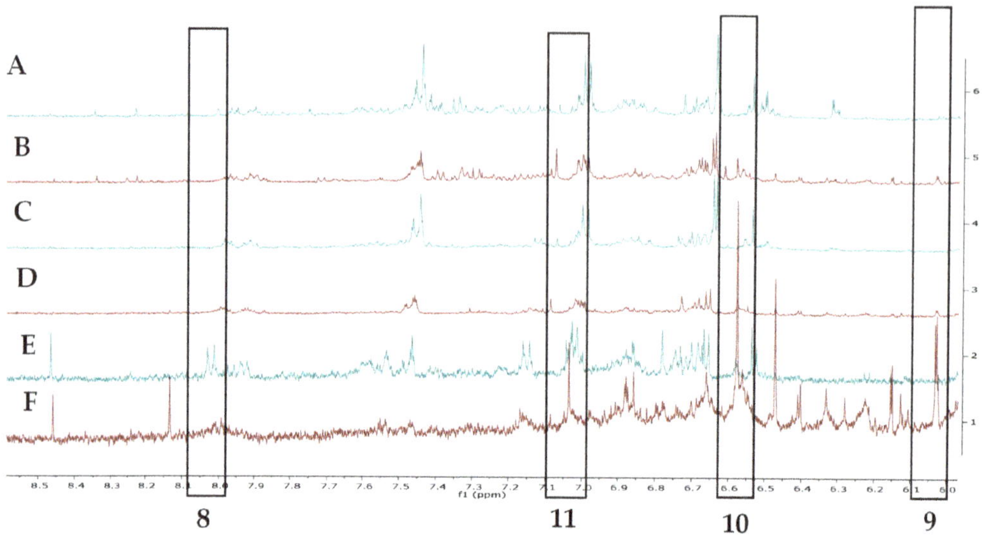

Figure 8. Comparison of ^{1}H-NMR spectra in the phenolics regions of Robinetta (**A**) and Charming Beauty plants grown in the field, (**A**,**B**) and (**D**), grown in the field and transferred to a climate chamber (**C**,**D**) and grown in a climate chamber (**E**,**F**). Metabolites associated with resistance *Western Flower Thrips* are labelled as kaempferol (8), epicatechin (9), epigallocatechin (10) and gallic acid (11).

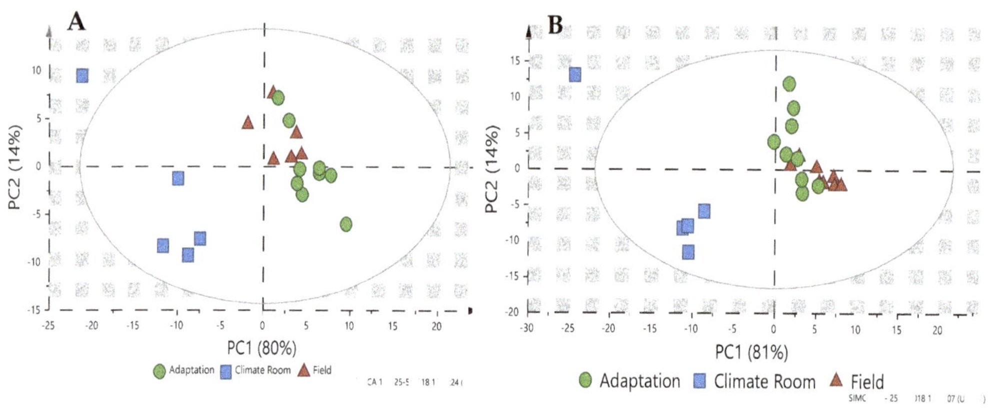

Figure 9. Score plot of PLS-DA based on Charming Beauty (**A**) and Robinetta (**B**) plants grown in the field (▲), plants grown in the field and transferred to a climate chamber (■) and plant grown in a climate room (•).

The important question one may ask is if there is a consistent difference between the two varieties independent of the environmental conditions. All compounds known to be associated with thrips resistance in *Gladiolus* [18] were higher in Robinetta, the resistant variety, for all three environmental conditions (Figure 10). Between environmental conditions there were some metabolomic differences, with a trend for triterpenoids to be lower under climate room conditions. A similar trend seemed to be present for the amino acids alanine, valine and threonine and sucrose (Figure 10). In contrast the concentrations of kaempferol were significantly higher when plants were grown in the climate room.

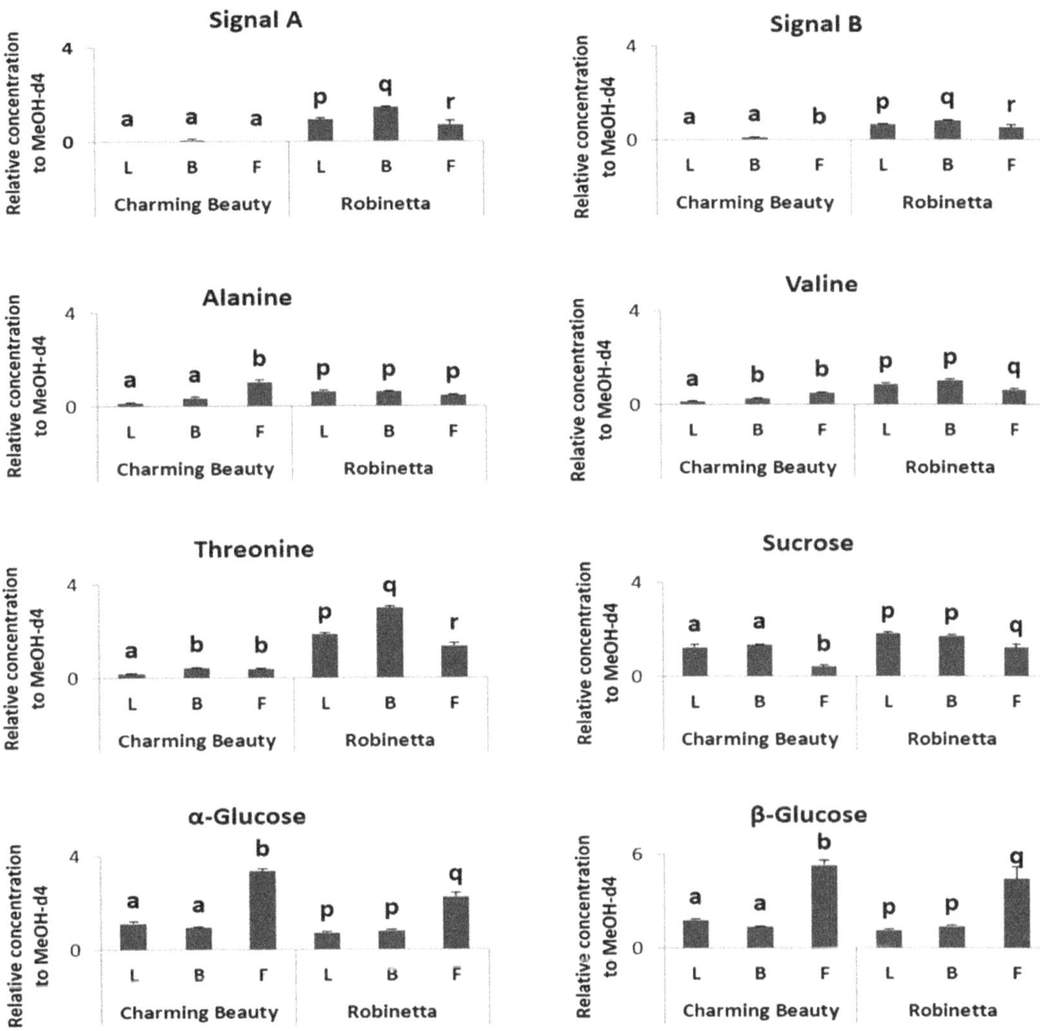

Figure 10. Relative concentrations, as proportions of the internal standard, in ^{1}HNMR spectra of signal A, signal B, alanine, valine, threonine, sucrose and kaempferol related to different environmental conditions of the thrips susceptible variety Charming Beauty (CB) and the resistant variety Robinetta (R). Data present the mean of four to six replicates ± SE of the mean. Signal A, signal B, threonine and kaempferol were analyzed by Kruskal-Wallis test. Alanine, valine and sucrose were analyzed by one-way ANOVA. Different letters refer to significant differences between varieties in each environmental condition at the 0.05 level.

All compounds known to be associated with susceptibility to thrips [18] were higher in Charming Beauty, the susceptible variety, for all three environmental conditions (Figure 11). Concentrations of α-glucose and β-glucose were lower in the climate room whereas concentrations of gallic acid and epigallocatechin were higher while the concentration of epicatechin was not affected by environmental conditions.

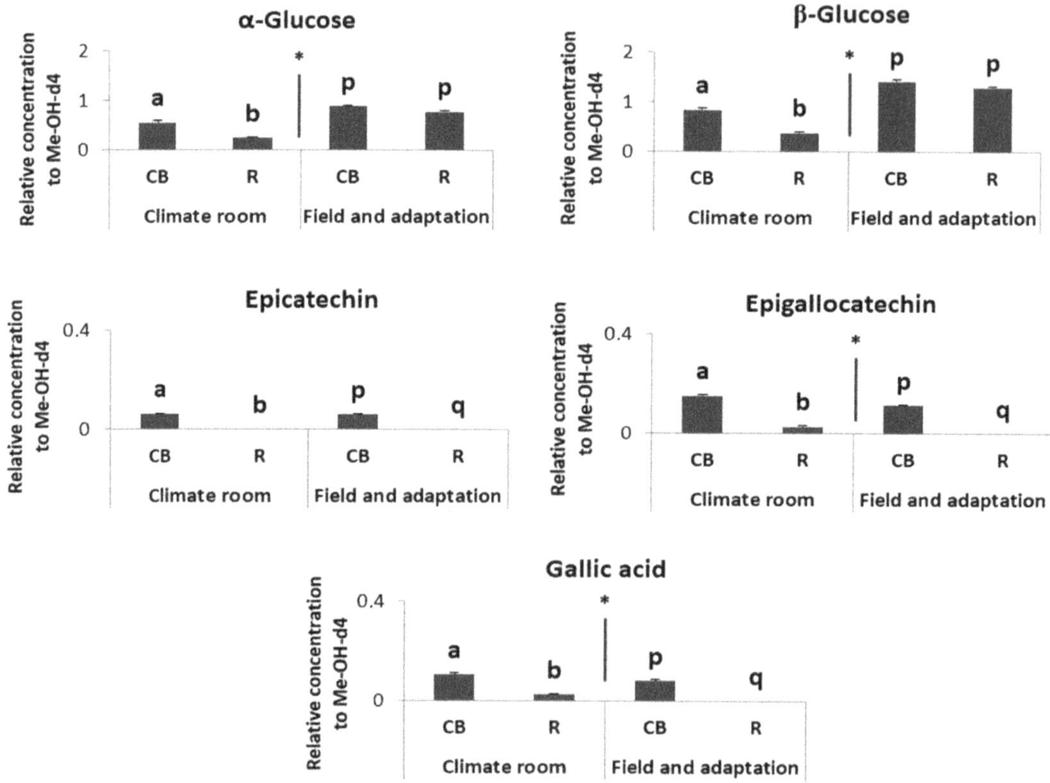

Figure 11. Relative concentrations, as proportions of the internal standard, in ^{1}HNMR spectra of α-glucose, β-glucose, epicatechin, epigallocatechin and gallic acid as the metabolites associated with the kaempferol related to different environmental conditions of the thrips susceptible variety Charming Beauty (CB) and the resistant variety Robinetta (R). Data present the mean of four to six replicates ± SE of the mean. α-glucose and β-glucose were analyzed by one-way ANOVA while epicatechin, epigallocatechin, gallic acid and kaempferol were analyzed by Kruskal-Wallis test. Different letters refer to significant differences between varieties in each environmental condition at the 0.05 level, while * indicate significant differences between environmental conditions at the 0.05 level.

Other metabolites that changed due to different environmental were luteolin and apigenin (Figure 12) as well as the organic acids formic- and malic acid (Figure 12). Concentrations of luteolin and apigenin were significantly higher in field grown plants while concentrations of formic- and malic acid were higher in the climate chamber.

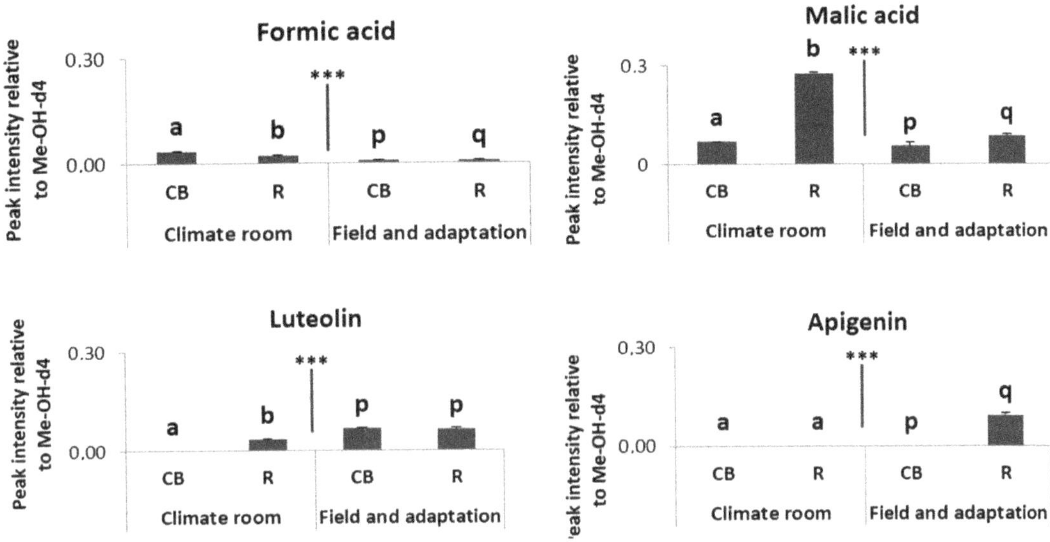

Figure 12. Relative concentrations, as proportions of the internal standard, in ^{1}HNMR spectra of formic acid, malic acid, luteolin and apigenin kaempferol related to different environmental conditions of the thrips susceptible variety Charming Beauty (CB) and the resistant variety Robinetta (R). Data present the mean of four to six replicates ± SE of the mean. Formic acid and malic acid were analyzed by one-way ANOVA while luteolin and apigenin were analyzed by Kruskal-Wallis tests. Different letters refer to significant differences between varieties in each environmental condition at the 0.05 level. *** Indicate significant differences between environmental conditions ($p < 0.000$).

3. Discussion

Robinetta and Charming Beauty showed consistent differences in WFT resistance over all development stages. Robinetta as the resistant variety exhibited more than 500-fold less silver damage at all plant development stages compared to Charming Beauty. Metabolomic profiles differed between the two varieties throughout all three plant stages. They revealed triterpenoid saponins and amino acids as metabolites associated with the resistant variety as shown in our earlier study [18]. Those compounds were consistently higher in Robinetta overall plant stages. Threonine was 10 times higher and triterpenoid saponins, valine and alanine were about five times higher in Robinetta. With the exception of valine all these compounds were observed to be negatively correlated with thrips resistance.

In Charming Beauty leaves were more damaged compared to buds and flowers: 50% of all damage occurred on the leaves. Metabolomic profiles differed among plant organs. Triterpenoid saponins were slightly higher in buds and amino acids were two to three times higher in buds and flowers compared to leaves. Patterns in metabolites related to resistance were, therefore, in line with patterns in silver damage. However, leaves represent a relatively larger area compared to buds and flowers so that differences in damage between organs may not solely be attributed to variation in metabolites. Although the silver damage on leaves was higher in the vegetative stage than in the two generative stages, we did not observe significant differences in leaf metabolites related to resistance (or to susceptibility) between leaves of different developmental stages. While in Robinetta damage was always much lower than in Charming Beauty, the concentrations of all compounds identified as being related to thrips resistance where much higher. In Robinetta, the relative concentrations of the triterpenoid saponins (signals A and B) of threonine, and of valine were much higher in leaves and buds than in flowers.

Whereas in many plants species old leaves are more attractive to WFT than young leaves we observed an opposite pattern in *Gladiolus* [13]. Damage to leaves was highest in the vegetative life-stage when leaves were on average young. However, vegetative and generative plant stages have similar leaf numbers while leaf area expands with age. Moreover, the concentration of defence compounds in leaves did not drop during successive life-stages. Having a higher concentration of defence compounds in buds and flowers is a way to protect the most valuable organs with respect to plant fitness from WFT. Similarly, Damle et al. [9] reported an accumulation of proteinase inhibitors in flowers as a protection against *Helicoverpa armigera* on tomato (*Lycopersicon esculentum* Mill). The pattern of damage across plant organs in Charming Beauty contrasted with the ornamental chrysanthemum, on which WFT preferred flowers over leaves. In the latter species WFT is attracted to pollen and it may find shelter in the flowers. In contrast to what we observed for Gladiolus, WFT caused more damage on plants with flower buds, than on plants with fully opened flowers or on plants with only leaves in *Impatiens walleriana* [10].

Differences in WFT resistance between the susceptible variety Charming Beauty and the resistant variety Robinetta remained constant across developmental stages. Furthermore, the concentrations of leaf metabolites identified to be associated with resistance in our earlier study [18] remained similar during the different development stages for both varieties. These results strongly suggest that markers for resistance in early developmental stages remain valid throughout the plant's life.

The effect of the environment on the metabolomic profile is clear between plants grown in the field and in the climate room, but the transition from the field into the climate chamber does not seem to cause many changes in the metabolome. Metabolites that were affected by the growing conditions included the flavonoids kaempferol, apigenin, and luteolin, as well as some organic acids: formic acid, gallic acid and malic acid. Climate room generally have a lower photosynthetic active radiation (PAR) level and UV-B dose compared to field conditions [19]. In the present study, light in the climate chamber was lower than in field conditions which might have caused the chemical variation. Kaempferol was at higher levels in the climate room grown plants. This is in accordance with the results reported by Muller et al. [20] for the perennial semi-aquatic plant *Hydrocotyle leucocephala* showing higher kaempferol concentrations for plants grown in climate room compared to plants grown in natural light conditions in the field. In contrast, luteolin and apigenin, were higher in field and field transition-grown plants. Markham et al. [21], reported that in the thallus of the common liverwort, *Marchantia polymorpha* the flavonoids, luteoline and apigenin, had a strong positive correlation to UV-B levels. Formic acid, gallic acid and malic acid were higher in climate room-grown plants whereas Jankanpaa et al. [2] reported that malic acid was more abundant in high-light plants than in low-light plants of *Arabidopsis*.

Concentrations of metabolites previously found to be related to thrips resistance were similar in each of the three environments while differences between the two varieties remained. Consequently, the environment seemed no to have affected the compounds related to constitutive thrips resistance in *Gladiolus*. In other words, resistance in *Gladiolus* seems mainly genetically determined.

Unlike secondary metabolites, amino acids belong to the primary metabolites and are part of the plants primary metabolism which is responsible for plant growth and development. Amino acids were reported by Jankanpaa et al. [2] as light-intensity dependent compounds in *Arabidopsis thaliana*. Valine was strikingly higher in plants grown under low light (30 μmol photons m^{-2} s^{-1}) conditions, alanine had higher concentrations in high light (600 μmol photons m^{-2} s^{-1}) and normal light (300 μmol photons m^{-2} s^{-1}). Threonine had accumulated in Arabidopsis one hour after transfer from a growth chamber into the field. In the present study, alanine, valine and threonine were slightly lower in the climate chamber with lower light intensity (Figure 11).

All together our results show that differences in plant defence compounds related to thrips resistance between a resistant and a susceptible variety persist during plant development and under different growing conditions. Therefore, they seem useful for breeding programs targeted at resistance. However, when breeding for resistance it is important not to impair bulb or flower production. These metabolites associated with resistance are among the most expensive defence metabolites (triterpenoid saponins) for plants to synthesize [22]. Thus, the higher expenditure in resistance may be one of the factors leading to a smaller dry mass of Robinetta compared to Charming Beauty [18]. More research on the costs of resistance would be needed for a successful breeding program.

4. Materials and Methods

4.1. Plant Materials

Two *Gladiolus nanus* varieties, (Charming Beauty and Robinetta), from vegetative, generative with buds and generative with flowers stages were obtained from the *Gladiolus* breeder Gebr. P. & M. Hermans (Lisse, The Netherlands).

4.2. Plant Development Stages

We grew plants outdoors in a field at Lisse, the Netherlands, to mimic the natural growing conditions. Plants at three development stages, i.e., vegetative, generative with buds and generative with flowers were collected from the field by carefully digging out the plants with their root system. Consequently, they were then potted and placed in a climate room (L:D, 18:6, 20 °C) for 7 days of further growth before they were infested by thrips. Robinetta was planted in the field 25 days earlier as Charming Beauty on May 2013. Because we harvested all the plants in a particular stage at the same day Robinetta plants had been in the field for a longer time period. Vegetative plants of Charming Beauty and Robinetta were thus collected after 65- and 90-days growth in the field, respectively. Plants with buds that just started to develop were collected after 75 days and 100 days in the field, respectively and plants with fully developed buds that started to open flowers were collected after 85 and 110 days, respectively. After collecting, plants were transferred to a climate chamber. Four to six replicates of all development stages were used for NMR metabolomics.

4.3. Different Environmental Conditions

Vegetative plants were grown under three different conditions: field, field transition and climate chamber. These plants were planted as bulbs to 9 × 9 cm pots filled with a 1:1 mixture of potting soil and dune sand. They were randomly placed in a climate chamber (L:D, 18:6, 20 °C, 70% relative humidity and 90–120 μmol photons m^{-2} s^{-1}) and grown for 70 days. Field-grown plants were planted and grown for 65 days (Charming Beauty) and 90 days (Robinetta). Part of these were carefully dug out from the field and transferred immediately into a climate room for 7 days. Plants from all conditions were harvested at the vegetative stage. Four to six replicates of all three conditions were used for NMR metabolomics.

4.4. Thrips Whole Plant Bioassay

For each of the two varieties, three to four plants per developmental stage were tested in a non-choice whole plant bioassay. Each plant was placed individually in a WFT proof cage, consisting of a plastic cylinder (80 cm height, 20 cm diameter), closed with a displaceable ring of WFT proof gauze. The cages were arranged in a fully randomized design. Two adult males and 18 adult females of western flower WFT were released in each cage and left for 10 days. Thereafter, silver damage, expressed as the leaf area damaged in mm^2, was visually scored for each plant. Silver damage in the buds and flowers in flowering plants were counted in mm^2 [12].

We calculated total damage per plant as the sum of the silver damage in all plant organs present in a certain stage. Because WFT damage in Robinetta was zero in many samples we could not use a two-way ANOVA to test for the effects of variety and developmental stage on silver damage. Instead, we tested for the effects of developmental stage for each variety separately. We used the Kruskal-Wallis test to do so for Robinetta and we used one-way ANOVA for Charming Beauty. Differences in total damage between the two varieties were analyzed by using a Mann-Whitney U test.

4.5. Metabolomic Profiling

4.5.1. Extraction of Plant Materials for NMR Metabolomics

The dried plant material was used to test for differences among leaves of the three developmental stages and for differences among buds and flowers in flowering plants for the two varieties using the standard protocol of sample preparation and ^{1}HNMR profiling [23].

Samples of 30 mg freeze-dried plant material were weighed into a 2 mL microtube and extracted with 1.5 mL of a mixture of phosphate buffer (pH 6.0) in deuterium oxide containing 0.05% trimethylsilylproprionic acid sodium salt-d_4 (TMSP) and methanol-d_4 (1:1). Samples were vortexed at room temperature for 1 min, ultrasonicated for 20 min and centrifuged at 13,000× g rpm for 10 min. an aliquot of 0.8 mL of the supernatant were transferred to 5 mm NMR tubes for ^{1}HNMR measurement.

4.5.2. NMR Analysis

^{1}HNMR spectra were recorded with a 500 MHz Bruker DMX-500 spectrometer (Bruker, Karlsruhe, Germany) operating at a proton NMR frequency of 500.13 MHz Deuterated methanol was used as the internal lock. Each ^{1}HNMR spectrum consisted of 128 scans requiring 10 min and 26 s acquisition time with following parameters: 0.16 Hz/point, pulse width (PW) of 30 (11.3 µs), and relaxation delay (RD) of 1.5 s. A pre-saturation sequence was used to suppress the residual water signal with low power selective irradiation at the water frequency during the recycle delay. Free induction decay (FIDs) was Fourier transformed with a line broadening (LB) of 0.3 Hz. The resulting spectra were manually phased and baseline corrected to the internal standard TMSP at 0.00 ppm, using TOPSPIN (version 3.5, Bruker). Two-dimensional J-resolved NMR spectra were acquired using 8 scans per 128 increments for F1 and 8 k for F2 using spectral widths of 5000 Hz in F2 (chemical shift axis) and 66 Hz in F1 (spin-spin coupling constant axis). Both dimensions were multiplied by sine-bell functions (SSB = 0) prior to double complex Fourier transformation. J-resolved spectra were tilted by 45 o, symmetrized about F1, and then calibrated to TMSP, using XWIN NMR (version 3.5, Bruker). 1H-1H correlated COSY spectra were acquired with a 1.0 s relaxation delay and 6361 Hz spectral width in both dimensions. The window function for the COSY spectra was Qsine (SSB = 0).

4.5.3. Data Processing

Spectral intensities were scaled to total intensity and reduced to integrated equal width (0.04 ppm) for the region of δ 0.32–10.0. The regions of δ 4.7–5.0 and δ 3.30–3.34 were excluded from analysis due to the presence of the residual signals of water and methanol. ^{1}HNMR spectra were automatically binned by AMIX software (version 3.7, Biospin, Bruker). Plant development stages data were further analyzed with principal component analysis (PCA) performed with SIMCA-P software (version 15.0 Umetrics, Umea, Sweden). Pareto scaling was used for PCA analysis. With the PCA we tested for differences in metabolomics profiles between the two varieties. Besides, different environmental conditions data were further analyzed with partial least square-discriminant analysis (PLS-DA) which used unit variance scaling.

The peak area of triterpenoid saponins at δ 1.28 and 0.92 ppm earlier reported to be related to trips resistance in *Gladiolus* [18] were close to zero in all plant development stages in Charming Beauty. We, therefore, analyzed differences in these signals with the Kruskal-Wallis test, while all others were analyzed by one-way ANOVA. The relative concentrations of threonine, valine, alanine, sucrose, α-glucose and β-glucose were ln-transformed to fit a normal distribution. For leaves, differences between the two varieties in the peak areas of triterpenoid saponins were analyzed with a Kruskal-Wallis test, while differences between the two varieties in other metabolites were analyzed with one-way ANOVA. Differences in relative concentrations of triterpenoid saponins between plant organs were analyzed with a Kruskal-Wallis test while differences in other metabolites were analyzed with one-way ANOVA within variety. Differences in the relative concentrations of compounds between leaves at different developmental stages were analyzed with one-way ANOVA within variety. Data were subsequently analyzed with the Scheffe post-hoc test. Differences in metabolite concentrations between plants grown under different conditions, were analyzed separately using one-way ANOVA. Data was log-transformed to fit a normal distribution. Triterpenoid saponins, threonine and kaempferol were analyzed by Kruskal-Wallis tests.

5. Conclusions

In the present study metabolomic profiling is able to differ between the two varieties throughout all three plant stages. Differences in resistance between the susceptible variety Charming Beauty and the resistant variety Robinetta remained constant across developmental stages. Furthermore, we found no differences in resistance of leaves among developmental stages for both varieties which was accompanied by the absence of differences among developmental stages of metabolites in leaves that were identified as associated with resistance. Together, these results strongly suggest that markers for resistance in early developmental stages remain valid throughout the plant's life.

Author Contributions: P.G.L.K. and K.A.L. designed the study. D.S.C.W. implemented the bio-assay experiments and sampling, carried out data analysis and interpretation. D.S.C.W. and Y.H.C. performed the NMR analysis. D.S.C.W. prepared the manuscript, which was commented on by P.G.L.K., K.A.L. and Y.H.C. All authors have read and agreed to the published version of the manuscript.

Funding: The first author held a grant of the Directorate General of Higher Education (DHGE) of the Republic of Indonesia to do the study.

Data Availability Statement: ^{1}H NMR data of the samples are deposited in the data storage of Natural Products Laboratory (Institute of Biology, Leiden University, Leiden, The Netherlands). The data can be provided upon request.

Acknowledgments: We thank the Dutch *Gladiolus* breeders Gebr. Hermans and VWS B.V. (Alkmaar, The Netherlands) for providing the different *Gladiolus* varieties. Suzanne Kos, Rita Rakhmawati and Mariá José Rodriguez-Lopez from Plant Sciences and Natural Products, Institute of Biology (IBL), Leiden University are thanked for their technical assistance.

Conflicts of Interest: The authors declare no conflict of interest.

References

1. Szakiel, A.; Pączkowski, C.; Henry, M. Influence of environmental abiotic factors on the content of saponins in plants. *Phytochem. Rev.* **2011**, *10*, 471–491. [CrossRef]
2. Jänkänpää, H.J.; Mishra, Y.; Schröder, W.P.; Jansson, S. Metabolic profiling reveals metabolic shifts in Arabidopsis plants grown under different light conditions. *Plant Cell Environ.* **2012**, *35*, 1824–1836. [CrossRef] [PubMed]
3. Rani, P.U.; Prasannalaxmi, K. Water stress induced physiological and biochemical changes in *Piper betle* L. and *Ricinus communis* L. plants and their effects on Spodoptera litura. *Allelopath. J.* **2014**, *33*, 25–41.
4. Boege, K.; Marquis, R.J. Facing herbivory as you grow up: The ontogeny of resistance in plants. *Trends Ecol. Evol.* **2005**, *20*, 441–448. [CrossRef] [PubMed]
5. Barton, K.E.; Koricheva, J. The Ontogeny of Plant Defense and Herbivory: Characterizing General Patterns Using Meta-Analysis. *Am. Nat.* **2010**, *175*, 481–493. [CrossRef] [PubMed]
6. De Jong, T.J.; Van Der Meijden, E. On the correlation between allocation to defence and regrowth in plants. *Oikos* **2000**, *88*, 503–508. [CrossRef]
7. Sun, H.H.; Wang, L.; Zhang, B.Q.; Ma, J.H.; Hettenhausen, C.; Cao, G.Y.; Sun, G.; Wu, J.; Wu, J. Scopoletin is a phytoalexin against *Alternaria alternata* in wild tobacco dependent on jasmonate signalling. *J. Exp. Bot.* **2014**, *65*, 4305–4315. [CrossRef]
8. Van Dam, N.M.; Horn, M.; Mareš, M.; Baldwin, I.T. Ontogeny constrains systemic protease inhibitor response in Nicotiana attenuata. *J. Chem. Ecol.* **2001**, *27*, 547–568. [CrossRef]
9. Damle, M.S.; Giri, A.P.; Sainani, M.N.; Gupta, V.S. Higher accumulation of proteinase inhibitors in flowers than leaves and fruits as a possible basis for differential feeding preference of Helicoverpa armigera on tomato (*Lycopersicon esculentum* Mill, Cv. Dhanashree). *Phytochemistry* **2005**, *66*, 2659–2667. [CrossRef]
10. Ugine, T.A.; Sanderson, J.P.; Wraight, S.P. Within-plant and temporal distribution of nymphal and adult western flower thrips, *Frankliniella occidentalis* (Thysanoptera: Thripidae), on flowers and foliage of greenhouse impatiens, Impatiens wallerana, and implications for pest population sampling. *Environ. Entomol.* **2006**, *35*, 507–515. [CrossRef]
11. Kirk, W.D.J. Aggregation and mating of thrips in flowers of *Calystegia sepium*. *Ecol. Entomol.* **1985**, *10*, 433–440. [CrossRef]

12. Mirnezhad, M.; Romero-González, R.R.; Leiss, K.A.; Choi, Y.H.; Verpoorte, R.; Klinkhamer, P.G.L. Metabolomic analysis of host plant resistance to thrips in wild and cultivated tomatoes. *Phytochem. Anal.* **2010**, *21*, 110–117. [CrossRef] [PubMed]
13. Leiss, K.A.; Choi, Y.H.; Abdel-Farid, I.B.; Verpoorte, R.; Klinkhamer, P.G.L. NMR metabolomics of thrips (*Frankliniella occidentalis*) resistance in Senecio hybrids. *J. Chem. Ecol.* **2009**, *35*, 219–229. [CrossRef] [PubMed]
14. Kirk, W.D.J.; Terry, L.I. The spread of the western flower thrips *Frankliniella occidentalis* (Pergande). *Agric. For. Entomol.* **2003**, *5*, 301–310. [CrossRef]
15. Jensen, S.E. Insecticide resistance in the western flower thrips, *Frankliniella occidentalis*. *Integr. Pest Manag. Rev.* **2000**, *5*, 131–146. [CrossRef]
16. Buitenhuis, R.; Shipp, J.L. Influence of plant species and plant growth stage on *Frankliniella occidentalis* pupation behaviour in greenhouse ornamentals. *J. Appl. Entomol.* **2008**, *132*, 86–88. [CrossRef]
17. Ullman, D.E.; Sherwood, J.L.; German, T.L. Thrips as vectors of plant pathogens. In *Thrips as Crop Pests*; Cabi Publishing: Oxfordshire, UK, 1997; pp. 539–565.
18. Wahyuni, D.S.C.; Young, H.C.; Leiss, K.; Klinkhamer, P.G.L. Morphological and Chemical Factors Related to Western Flower Thrips Resistance in the Ornamental Gladiolus. *Plants* **2021**, *10*, 3390. [CrossRef]
19. Deckmyn, G.; Impens, I. The ratio UV-B/photosynthetically active radiation (PAR) determines the sensitivity of rye to increased UV-B radiation. *Environ. Exp. Bot.* **1997**, *37*, 3–12. [CrossRef]
20. Muller, V.; Lankes, C.; Albert, A.; Winkler, J.B.; Zimmermann, B.F.; Noga, G.; Hunsche, M. Concentration of hinokinin, phenolic acids and flavonols in leaves and stems of *Hydrocotyle leucocephala* is differently influenced by PAR and ecologically relevant UV-B level. *J. Plant Physiol.* **2015**, *173*, 105–115. [CrossRef]
21. Markham, K.R.; Ryan, K.G.; Bloor, S.J.; Mitchell, K.A. An increase in the luteolin: Apigenin ratio in *Marchantia polymorpha* on UV-B enhancement. *Phytochemistry* **1998**, *48*, 791–794. [CrossRef]
22. Gershenzon, J. Metabolic costs of terpenoid accumulation in higher plants. *J. Chem. Ecol.* **1994**, *20*, 1281–1328. [CrossRef]
23. Kim, H.K.; Wilson, E.G.; Choi, Y.H.; Verpoorte, R. Metabolomics: A tool for anticancer lead-finding from natural products. *Plant. Med.* **2010**, *76*, 1094–1102. [CrossRef]

plants

MDPI

Article

Identification of Daphnane Diterpenoids from *Wikstroemia indica* Using Liquid Chromatography with Tandem Mass Spectrometry

Mi Zhang [1], Kouharu Otsuki [1,*], Reo Takahashi [1], Takashi Kikuchi [1], Di Zhou [2], Ning Li [2] and Wei Li [1,*]

1 Faculty of Pharmaceutical Sciences, Toho University, Miyama 2-2-1, Funabashi 274-8510, Chiba, Japan; zhangmi495@gmail.com (M.Z.); takashi.kikuchi@phar.toho-u.ac.jp (T.K.)

2 Key Laboratory of Innovative Traditional Chinese Medicine for Major Chronic Diseases of Liaoning Province, Key Laboratory for TCM Material Basis Study and Innovative Drug Development of Shenyang City, School of Traditional Chinese Materia Medica, Shenyang Pharmaceutical University, Shenyang 110016, China; zd930322@126.com (D.Z.); liningsypharm@163.com (N.L.)

* Correspondence: kouharu.otsuki@phar.toho-u.ac.jp (K.O.); liwei@phar.toho-u.ac.jp (W.L.); Tel.: +81-47-4721396 (K.O.); Tel.: +81-47-4721161 (W.L.)

Abstract: Liquid chromatography coupled with tandem mass spectrometry (LC-MS/MS) has emerged as a powerful tool for the rapid identification of compounds within natural resources. Daphnane diterpenoids, a class of natural compounds predominantly found in plants belonging to the Thymelaeaceae and Euphorbiaceae families, have attracted much attention due to their remarkable anticancer and anti-HIV activities. In the present study, the presence of daphnane diterpenoids in *Wikstroemia indica*, a plant belonging to the Thymelaeaceae family, was investigated by LC-MS/MS analysis. As a result, 21 daphnane diterpenoids (**1–21**) in the stems of *W. indica* were detected. Among these, six major compounds (**12**, **15**, **17**, **18**, **20**, and **21**) were isolated and their structures were unequivocally identified through a comprehensive analysis of the MS and NMR data. For the minor compounds (**1–11**, **13**, **14**, **16**, and **19**), their structures were elucidated by in-depth MS/MS fragmentation analysis. This study represents the first disclosure of structurally diverse daphnane diterpenoids in *W. indica*, significantly contributing to our understanding of bioactive diterpenoids in plants within the Thymelaeaceae family.

Keywords: daphnane diterpenoids; LC-MS/MS; *Wikstroemia indica*; MS/MS fragmentation

1. Introduction

Liquid chromatography coupled with high-resolution tandem mass spectrometry (LC-HR-MS/MS), usually equipped with an electrospray ionization source, has high adaptability across a broad spectrum of compounds, offering high mass accuracy and sensitivity. Moreover, it provides information-rich fragmentation through product ion spectra, thereby potentially revealing details about the molecular formula and structure of diverse secondary metabolites found in plants [1]. The conventional phytochemical research process often necessitates substantial amounts of accessible plant materials and time-consuming purification procedures, whereas applying LC-MS/MS analysis on crude plant extracts at the early stage of phytochemical investigations allows for the rapid identification of the compounds [2,3].

Daphnane diterpenoids, characterized by their *trans*-fused 5/7/6-tricyclic skeleton, have garnered attention for their diverse biological activities, including anticancer [4], anti-HIV [5], analgesic [6], anti-inflammatory [7], and neurotrophic activities [8,9]. These diterpenoids are predominantly found in plants of the Thymelaeaceae and Euphorbiaceae families, with the majority of them sourced from the Thymelaeaceae family [10]. Previous phytochemical investigations on plants of the Thymelaeaceae family have reported the isolation of daphnane diterpenoids from 16 genera, such as *Daphne*, *Pimelea*, *Stellera*, and *Wikstroemia*. Among these, the *Wikstroemia* genus, comprising over 70 species, holds significant potential as a source of daphnane diterpenoids. Isolation of daphnane diterpenoids has hitherto been reported from a number of species, including *W. monticola* [11], *W. mekongenia* [12], *W. retusa* [13,14], *W. polyantha* [15], *W. chamaedaphne* [16–18], *W. chuii* [19], and *W. ligustrina* [20]. It is evident that the *Wikstroemia* genus remains relatively underexplored in the research into daphnane diterpenoids.

Wikstroemia indica (L.) C. A. Mey. is a semi-evergreen shrub mainly distributed in southeastern China, which has long been used as a traditional Chinese medicine for the treatment of bronchitis, hepatitis, and cancer [21]. Recent studies have revealed that the extract of this plant exhibited antiallergic [22], anti-inflammatory [23], and antineoplastic properties [24], therefore heightening interest in its pharmacological exploration. While previous phytochemical investigations of *W. indica* have yielded coumarins [25], flavonoids [26], lignans [27], and sesquiterpenoids [28], the presence of daphnane diterpenoids has yet to be documented.

During our ongoing research aimed at discovering biological diterpenoids from plants of the Thymelaeaceae family [5,20,29,30], this study comprehensively examined and identified daphnane diterpenoids in the stems of *W. indica* using LC-MS/MS analysis.

2. Results and Discussion

2.1. Detection of Daphnane Diterpenoids in W. indica by LC-MS/MS

Due to the limited availability of plant material, the presence of daphnane diterpenoids in *W. indica* was initially examined by LC-MS/MS analysis. The criteria for validating that the detected peaks represented daphnane diterpenoids were established based on a synthesis of our previous studies and literature review [20,31,32]. These criteria included: (1) In the mass spectra, protonated molecular ions ($[M + H]^+$) and/or ammonium adduct ions ($[M + NH_4]^+$) were observed in positive ion mode, while deprotonated molecular ions ($[M–H]^-$) and/or formate adduct ions ($[M + HCOO]^-$) were observed in negative ion mode. (2) In the product ion spectrum obtained from the protonated molecular ion as a precursor ion, a diagnostic ion at *m/z* 253 ($C_{17}H_{17}O_2$) or 269 ($C_{17}H_{17}O_3$) was observed in the positive ion mode [31]. (3) The characteristic C_{17} product ions derived from C_{20} skeletons with the neutral loss of $C_3H_4O_2$ were observed [32]. (4) When the ion peaks originated from a macrocyclic daphnane orthoester (MDO), the second and third criteria were not applicable. Instead, product ion peaks derived from continuous losses of H_2O and CO were observed at the mass range of *m/z* 250–350 and *m/z* 400–550, respectively [20].

To enhance the detecting sensitivity, a crude diterpenoid fraction was prepared from the 95% EtOH extract using a sequence of procedures, including EtOAc–H_2O partition and Diaion HP-20 column chromatography. Subsequent LC-MS/MS analysis of the crude diterpenoid fraction, guided by the aforementioned criteria, resulted in the detection of three major daphnane diterpenoid peaks (**15**, **20**, and **21**), strongly suggesting the occurrence of daphnane diterpenoids in *W. indica* stems. It was noteworthy that detecting daphnane diterpenoids can be challenging due to their chromatographic behavior, which was sometimes similar to common plant constituents, such as fatty acids, acylglycerols, and chlorophyll [33].

To further enhance the sensitivity of LC-MS/MS detection of daphnane diterpenoids, a portion of the crude diterpenoid fraction underwent additional fractionation through gradient HPLC. As a result, a total of 21 daphnane diterpenoid peaks (**1**–**21**) were detected from three out of twelve subfractions (Figure S1). Importantly, all these peaks were subsequently confirmed to be present in the crude diterpenoid fraction through extracted ion chromatogram (XIC) analysis (Figure 1, Table 1).

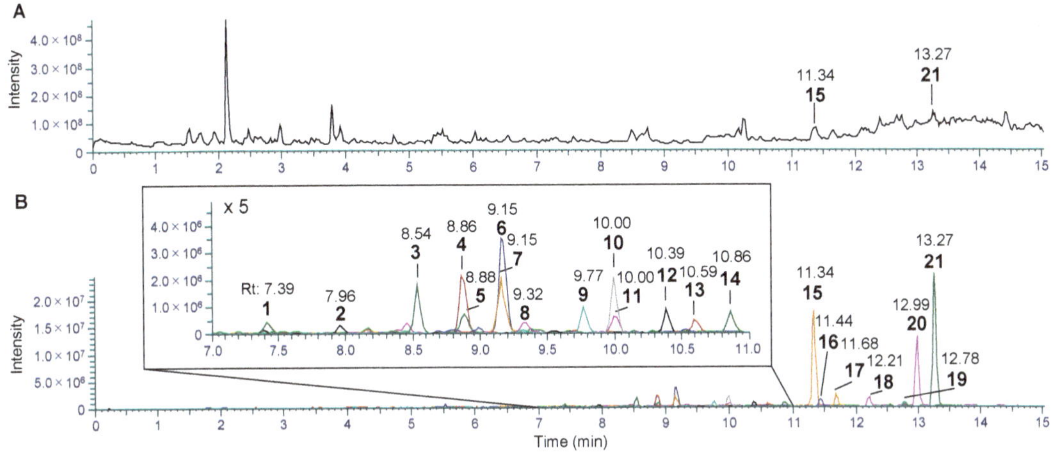

Figure 1. (**A**) Total ion chromatogram in the positive ion mode and (**B**) extracted ion chromatogram from the crude diterpenoid fraction of the stems of *W. indica*.

Table 1. Daphnane diterpenoids **1–21** identified from the stems of *W. indica*.

No.	Rt (min)	Molecular Formula	$[M + H]^+$ (*m*/*z*) Detected Mass (*m*/*z*)	Error (ppm)	ESI-MS/MS (*m*/*z*) (%) [c]	Identification
1 [a]	7.39	$C_{55}H_{62}O_{18}$	1011.3984	−2.97	793 (36), 775 (22), 765 (19), 747 (20), 731 (34), 703 (23), 689 (16), 671 (47), 659 (12), 653 (17), 643 (17), 625 (24), 609 (18), 567 (24), 549 (100), 531 (15), 521 (39), 507 (36), 503 (37), 493 (10), 489 (66), 479 (18), 477 (18), 471 (37), 461 (68), 459 (23), 443 (80), 433 (24), 431 (25), 425 (23), 415 (31), 403 (27), 397 (19), 375 (22), 363 (18), 339 (23), 325 (16), 307 (37), 295 (22), 291 (21), 279 (37), 263 (22), 221 (21), 183 (18), 181 (47), 153 (16), 141 (25), 105 (65)	daphneodorin C [29]
2 [a]	7.96	$C_{55}H_{62}O_{18}$	1011.3984	−2.43	793 (52), 775 (30), 765 (19), 747 (19), 731 (22), 707 (16), 689 (19), 671 (57), 653 (23), 643 (20), 629 (21), 625 (23), 611 (28), 593 (18), 583 (17), 549 (72), 531 (17), 521 (22), 507 (29), 503 (20), 489 (73), 479 (22), 471 (54), 461 (68), 453 (22), 443 (56), 425 (18), 415 (19), 375 (15), 363 (29), 307 (37), 291 (26), 279 (16), 221 (25), 181 (47), 163 (41), 141 (22), 105 (100)	daphneodorin B [29]
3 [a]	8.54	$C_{34}H_{50}O_9$	603.3527	0.36	585 (29), 361 (10), 343 (23), 325 (20), 315 (6), 313 (5), 307 (10), 297 (16), 295 (5), 279 (11), 271 (7), 269 (6), 267 (8), 253 (15), 207 (100), 203 (5), 107 (8), 95 (8), 81 (6)	wikstroelide M [14]

Table 1. *Cont.*

No.	Rt (min)	Molecular Formula	[M + H]+ (m/z) Detected Mass (m/z)	Error (ppm)	ESI-MS/MS (m/z) (%) [c]	Identification
4 [a]	8.86	$C_{30}H_{42}O_9$	547.2888	−2.45	529 (35), 511 (30), 501 (32), 493 (27), 483 (59), 467 (29), 465 (64), 449 (21), 447 (29), 439 (25), 437 (40), 423 (23), 421 (25), 419 (29), 405 (28), 395 (20), 341 (27), 323 (24), 295 (26), 283 (29), 255 (29), 239 (21), 236 (23), 235 (100), 233 (47), 227 (22), 215 (36), 203 (28), 199 (21), 193 (24), 187 (27), 161 (29), 135 (36), 133 (33)	pimelotide C [34]
5	8.88	$C_{37}H_{50}O_{10}$	655.3469	−1.24	619 (10), 515 (28), 497 (79), 479 (92), 469 (38), 467 (14), 461 (21), 451 (100), 443 (16), 439 (11), 433 (71), 423 (24), 421 (12), 415 (20), 405 (27), 367 (12), 311 (19), 293 (45), 275 (16), 265 (31), 263 (10), 251 (14), 247 (10), 225 (10), 211 (16), 133 (18), 123 (11), 105 (77)	kraussianin [35]
6 [a]	9.15	$C_{40}H_{46}O_{12}$	719.3062	−1.33	507 (3), 489 (4), 359 (4), 341 (12), 323 (18), 311 (3), 305 (4), 295 (12), 277 (4), 269 (8), 177 (75), 149 (95), 121 (11), 107 (100), 81(3)	acutilobin C [36]
7 [a]	9.15	$C_{30}H_{44}O_8$	533.3104	−1.00	515 (36), 497 (66), 479 (100), 469 (28), 467 (16), 461 (66), 451 (53), 449 (25), 443 (20), 433 (59),425 (24), 423 (13), 421 (16), 415(21), 407 (14), 405 (21), 403 (16), 309 (14), 291 (12), 281 (13), 263 (12), 211 (13), 187 (19), 185 (16), 159 (13), 135 (13), 133 (27)	pimelea factor S_6 [37]
8 [a]	9.32	$C_{44}H_{54}O_{12}$	775.3690	0.20	635 (15), 617 (21), 599 (22), 545 (63), 527 (33), 495 (47), 477 (85), 459 (100), 449 (21), 447 (28), 441 (37), 431 (50), 429 (20), 423 (90), 419 (18), 413 (29), 405 (24), 401 (23), 319 (18), 309 (24), 291 (24), 281 (18), 279 (23), 263 (19), 251 (18), 151 (27), 105 (88)	stelleralide H [38]
9 [a]	9.77	$C_{39}H_{46}O_{11}$	691.3133	−2.63	509 (4), 505 (2), 491 (5), 359 (7), 341 (20), 323 (21), 313 (4), 311 (3), 305 (3), 297 (2), 295 (13), 277 (4), 269 (9), 267 (3), 261 (3), 151 (100), 147 (75), 133 (3)	daphneodorin D [29]
10 [a]	10.00	$C_{40}H_{48}O_{12}$	721.3203	−1.60	509 (4), 491 (7), 359 (6), 341 (22), 323 (23), 313 (3), 311 (3), 305 (5), 295 (16), 277 (4), 269 (11), 267 (4), 261 (3), 177 (95), 151 (100), 95 (6), 81 (6)	acutilobin D [36]
11 [a]	10.00	$C_{44}H_{54}O_{12}$	775.3687	−0.09	563 (28), 545 (100), 513 (35), 495 (60), 477 (41), 467 (45), 465 (24), 449 (78), 441 (16), 437 (35), 431 (34), 425 (20), 423 (20), 421 (24), 419 (19), 391 (18), 309 (17), 291 (19), 263 (19), 105 (40)	gnidimacrin [39]
12 [b]	10.39	$C_{51}H_{58}O_{14}$	912.4147	−1.98	773 (48), 651 (47), 633 (58), 615 (34), 543 (100), 511 (57), 493 (82), 481 (20), 475 (58), 465 (54), 463 (36), 447 (90), 435 (43), 429 (38), 421 (47), 419 (21), 417 (22), 327 (23), 309 (40), 291 (26), 279 (22), 105 (55)	stelleralide G [38]

Table 1. *Cont.*

No.	Rt (min)	Molecular Formula	$[M + H]^+$ (m/z)		ESI-MS/MS (m/z) (%) [c]	Identification
			Detected Mass (m/z)	Error (ppm)		
13 [a]	10.59	$C_{37}H_{46}O_{11}$	667.3107	−0.93	545 (100), 527 (66), 509 (59), 499 (24), 491 (37), 483 (26), 481 (52), 465 (55), 463 (37), 453 (23), 447 (28), 445 (31), 435 (35), 419 (21), 417 (21), 357 (23), 321 (27), 295 (22), 235 (96), 231 (61), 203 (54), 185 (27), 173 (20), 153 (21), 105 (45)	stelleralide C [5]
14 [a]	10.86	$C_{39}H_{44}O_{10}$	673.2999	−1.27	359 (1), 341 (3), 323 (4), 295 (3), 277 (1), 269 (2), 149 (100), 131 (6), 107 (21)	12-*O*-(*E*)-cinnamoyl-9,13,14-ortho-(2*E*,4*E*,6*E*)-decatrienylidyne-5β,12β-dihydroxyresiniferonol-6α,7α-oxide [40]
15 [b]	11.34	$C_{30}H_{44}O_8$	533.3107	−0.23	361 (9), 343 (24), 325 (49), 307 (35), 297 (28), 279 (30), 267 (58), 253 (100), 203 (16), 155 (6)	simplexin [41]
16	11.44	$C_{32}H_{44}O_8$	557.3109	−0.08	361 (8), 343 (23), 325 (33), 313 (7), 307 (30), 297 (19), 295 (11), 285 (6), 279 (20), 277 (6), 267 (55), 261 (6), 253 (77), 251 (7), 249 (11), 225 (6), 179 (100)	wikstrotoxin B [11]
17 [b]	11.68	$C_{39}H_{46}O_{10}$	675.3151	−1.85	675 (9), 509 (2), 507 (2), 491 (2), 359 (5), 341 (15), 323 (21), 313 (4), 311 (3), 305 (4), 295 (13), 277 (5), 269 (11), 267 (3), 265 (2), 261 (2), 249 (2), 241 (2), 237 (2), 209 (2), 151 (100), 133 (4)	12-*O*-(*E*)-cinnamoyl-9,13,14-ortho-(2*E*,4*E*)-decadienylidyne-5β,12β-dihydroxyresiniferonol-6α,7α-oxide [40]
18 [a]	12.21	$C_{37}H_{50}O_9$	639.3527	−0.10	621 (12), 499 (30), 481 (93), 463 (100), 453 (32), 445 (35), 436 (23), 435 (57), 417 (34), 407 (15), 311 (5), 293 (16), 265 (23), 105 (67)	wikstromacrin [20]
19 [a]	12.78	$C_{36}H_{50}O_{10}$	643.3458	−2.97	365 (2), 359 (4), 341 (11), 323 (11), 313 (3), 305 (2), 295 (10), 277 (4), 269 (6), 267 (3), 207 (100), 189 (3), 107 (3)	wikstroelide A [13]
20 [b]	12.99	$C_{37}H_{50}O_9$	639.3525	−0.28	621 (16), 499 (28), 481 (73), 463 (100), 453 (31), 451 (12), 445 (55), 435 (55), 433 (14), 423 (12), 417 (34), 409 (12), 407 (10), 405 (16), 311 (13), 293 (24), 275 (12), 265 (24), 251 (12), 133 (12), 105 (99)	pimelea factor P_2 [41]
21 [b]	13.27	$C_{34}H_{48}O_8$	585.3409	−2.14	361 (14), 343 (20), 325 (41), 313 (10), 307 (25), 297 (23), 295 (13), 279 (25), 267 (59), 253 (99), 249 (16), 207 (100)	huratoxin [42]

[a] Identifications were confirmed with the daphnane diterpenoids isolated in our previous studies. [b] Isolated daphnane diterpenoids in this study. [c] ESI-MS/MS of $[M + H]^+$ ion for peaks **1–11** and **13–21** and ESI-MS/MS of $[M + NH_4]^+$ ion for peak **12**.

2.2. LC-MS Guided Isolation and Structural Determination of Major Daphnane Diterpenoids

An LC-MS guided isolation was carried out to obtain daphnane diterpenoids. The crude diterpenoid fraction was subjected to ODS column chromatography and eluted with a stepwise gradient of MeOH–H_2O. The fractions, in which daphnane diterpenoids were detected by LC-MS/MS analysis, were subjected to silica gel column chromatography and eluted with a gradient of *n*-hexane–EtOAc–MeOH–HCOOH. Those fractions containing daphnane diterpenoids were further purified by preparative HPLC and resulted in the isolation of six major daphnane diterpenoids (**12**, **15**, **17**, **18**, **20**, and **21**) (Figure 2).

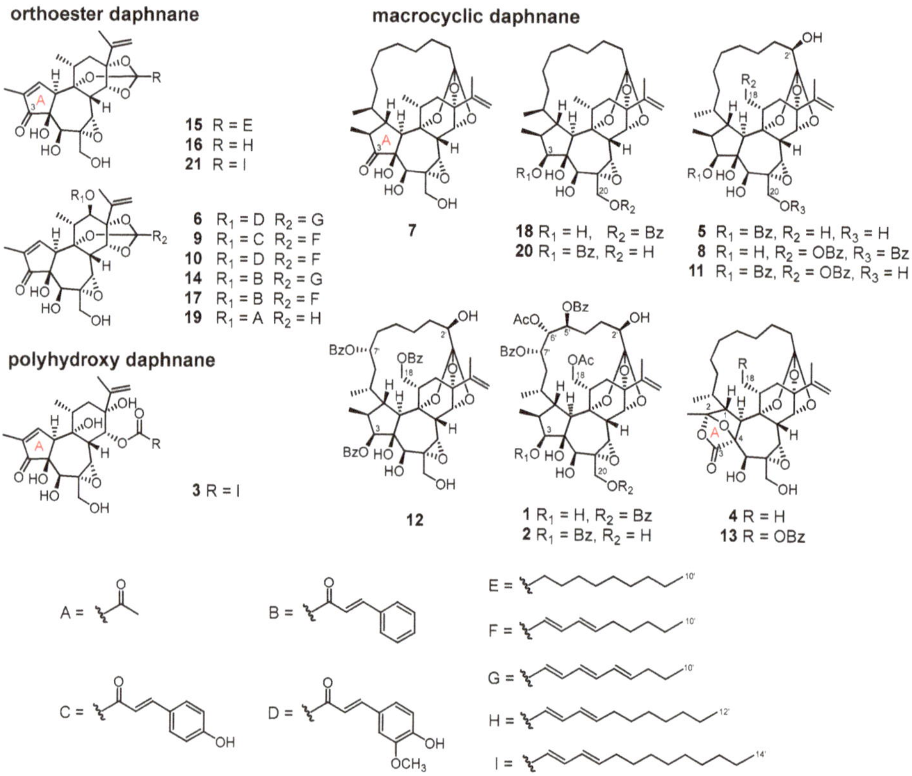

Figure 2. Structures of daphnane diterpenoids **1–21**.

The isolated compounds were identified by detailed NMR and MS spectroscopic analyses. In the ^{1}H- and ^{13}C-NMR spectra, the characteristic resonances for an isopropenyl moiety at δ_H 4.83–4.99 (H_a-16), 4.94–5.17 (H_b-16), and 1.72–1.84 (H_3-17), an epoxy group at δ_H 3.32–3.57 (H-7), δ_C 60.0–61.0 (C-6), and 63.5–64.4 (C-7), and an orthoester group at 116.5–119.8 (C-1′), indicated that they were daphnane diterpenoids (Tables S1 and S2) [10].

For compounds **15**, **17**, and **21**, the characteristic resonances of α,β-unsaturated carbonyl moiety at δ_H 7.56–7.61 (H-1), δ_C 160.5–161.4 (C-1), 136.6–136.9 (C-2), and 209.5–209.9 (C-3) indicated they belong to orthoester daphnane diterpenoids (Table S1). The presence of a decanoate moiety at C-1′ in **15** was deduced from the proton resonance for the aliphatic methylene multiplets at δ_H 1.23–1.93 and a terminal methyl triplet at δ_H 0.86 (H_3-10′). The 2*E*,4*E*-tetradecadienylidyne moiety of **21** was defined from the proton resonances for a conjugated diene at δ_H 5.89 (d, *J* = 15.5 Hz, H-2′), 6.68 (dd, *J* = 15.5, 10.6 Hz, H-3′), 6.04 (dd, *J* = 15.2, 10.6 Hz, H-4′), and 5.83 (dt, *J* = 15.2, 7.2 Hz, H-5′), and a *n*-nonyl moiety including eight methylenes at δ_H 1.24–2.07 and a terminal methyl group at δ_H 0.86 (t, *J* = 6.9 Hz, H_3-14′). The presence of cinnamoyloxy moiety of **17** was deduced from a *trans*-olefinic moiety at δ_H 6.35 (d, *J* = 15.9 Hz, H-2″) and 7.62 (d, *J* = 15.9 Hz, H-3″), and a phenyl moiety at δ_H 7.37–7.51 (each multiplet, H-5″ to H-9″), as well as the carbon resonance for an ester carbonyl at δ_C 165.8 (C-1″), and a 2*E*,4*E*-decadienylidyne moiety was confirmed by the proton resonances at δ_H 5.65 (d, *J* = 15.4 Hz, H-2′), 6.66 (dd, *J* = 15.4, 10.6 Hz, H-3′), 6.04 (dd, *J* = 15.1, 10.6 Hz, H-4′), and 5.85 (dt, *J* = 15.1, 7.0 Hz, H-5′), and a *n*-pentyl moiety including four methylenes at δ_H 2.08 (H-6′), 1.37 (H-7′), and 1.27 (H-8′, 9′), and a terminal methyl group at δ_H 0.87 (t, J = 6.8 Hz, H_3-10′). The location of the cinnamoyl moiety at C-12 was confirmed by the HMBC correlation from H-12 to C-1″. Thus, **15**, **17**, and **21** were determined as simplexin (**15**) [41], 12-*O*-(*E*)-cinnamoyl-9,13,14-ortho-(2*E*,4*E*)-decadienylidyne-5*β*,12*β*-dihydroxyresiniferonol-6*α*,7*α*-oxide (**17**) [40], and huratoxin (**21**) [42].

On the other hand, compounds **12**, **18**, and **20** belong to MDOs since the resonances of H-1 were observed as the methine proton at δ_H 2.11–2.97 and the methyl proton resonances at δ_H 0.79–1.26 of H_3-10′ were observed as doublet (Table S2) [10]. The presence of benzoyl moieties of **12**, **18**, and **20** were determined by the aromatic proton resonances at δ_H 8.02–8.16 (H-2′,6′), 7.35–7.47 (H-3′,5′), and 7.53–7.59 (H-4′). The locations of benzoyl moieties at C-3, C-18, and C-7′ of **12**, C-20 of **18**, and C-3 of **20** were confirmed by the HMBC experiment. Thus, **12**, **18**, and **20** were determined as stelleralide G (**12**) [38], wikstromacrin (**18**) [20], and pimelea factor P_2 (**20**) [41].

2.3. Identification of Minor Daphnane Diterpenoids by MS/MS Fragmentation Elucidation

To identify the minor daphnane diterpenoids (**1–11**, **13**, **14**, **16**, and **19**), which could not be isolated, MS/MS fragmentation elucidation was performed. These daphnane diterpenoids exhibited abundant ions in the product ion spectra derived from the protonated molecular ion as a precursor ion. Consequently, a detailed interpretation of the MS/MS fragmentation pathways in positive mode for these peaks was conducted (Figures 3 and 4). The identification of those peaks was confirmed by the LC-MS data which were in full accordance with the corresponding compounds isolated in our previous studies (Table 1) [5,29,30].

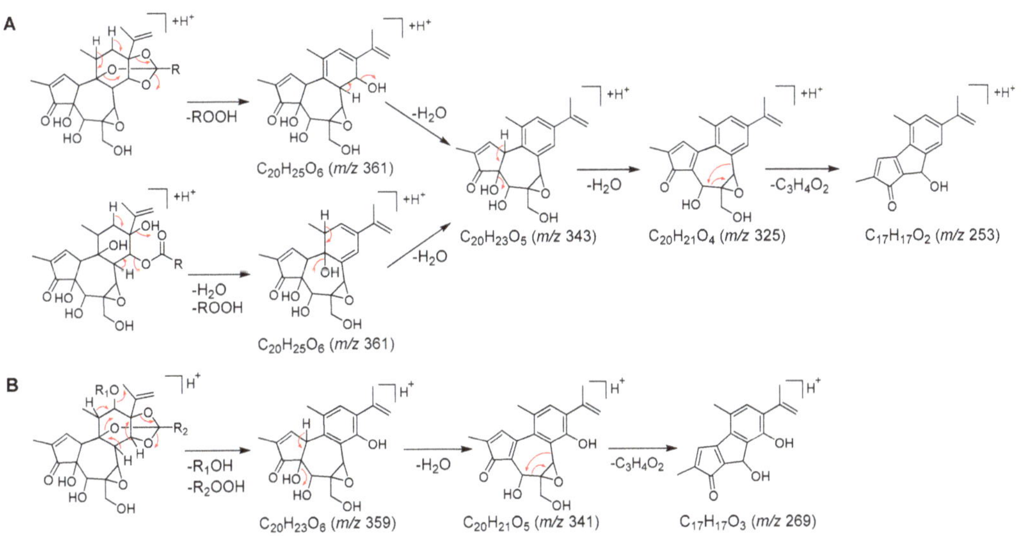

Figure 3. Proposed ESI-MS/MS fragmentation pathways for minor daphnane diterpenoids in positive mode. (**A**) peaks **3** and **16**, and (**B**) peaks **6**, **9**, **10**, **14**, and **19**.

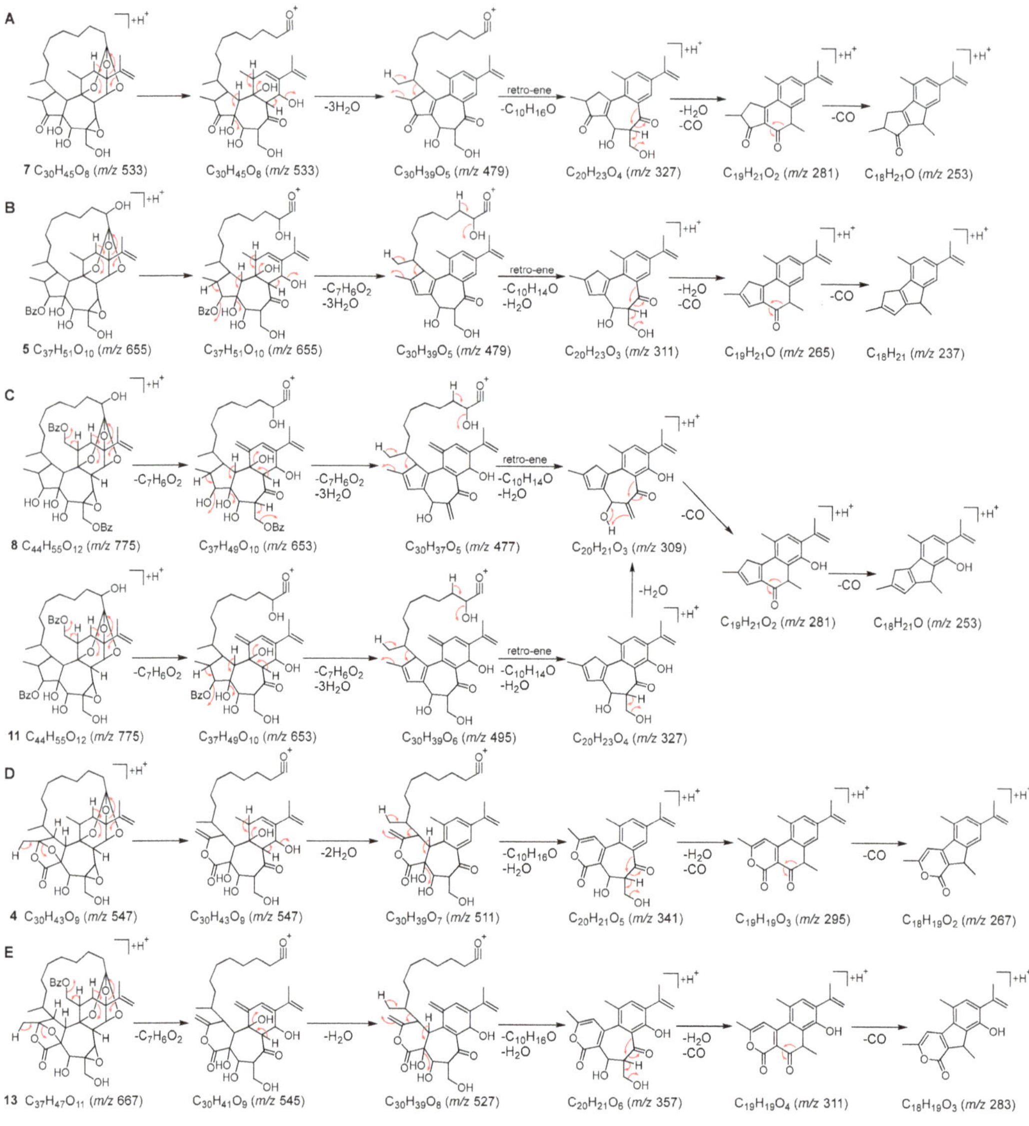

Figure 4. Proposed ESI-MS/MS fragmentation pathways for minor daphnane diterpenoids in positive mode. (**A**) peak **7**, (**B**) peak **5**, (**C**) peaks **8**, **11**, (**D**) peak **4**, and (**E**) peak **13**.

In the product ion spectra of peaks **3** and **16**, the characteristic product ion was observed at *m*/*z* 253 ($C_{17}H_{17}O_2$), which was produced by the loss of the 6,7-epoxy moiety, along with the oxymethylene at C-20, as a $C_3H_4O_2$ unit due to cleavage occurring at the B-ring. This observation suggested that both **3** and **16** were daphnane diterpenoids lacking a substituent at C-12 (Figures 3A and S22). Furthermore, the product ions at *m*/*z* 207 ($C_{14}H_{23}O$), 95 (C_6H_7O), and 81 (C_5H_5O) for peak **3**, and at *m*/*z* 179 ($C_{12}H_{19}O$), 95 (C_6H_7O), and 81 (C_5H_5O) for peak **16** indicated that a 2*E*,4*E*-tetradecadienoyl moiety was ester-linked to the C-ring in peak **3** and a 2*E*,4*E*-dodecadienoyl moiety was ester-linked in peak **16**. However, the molecular formula of peak **3** was 18 Da (H_2O) larger than that of the orthoester daphnane, huratoxin (**21**) [42] and the fragment ion of $[M + H–H_2O]^+$ appeared with greater intensity in the mass spectrum of peak **3** (Figure S23). Based on these observations, it was concluded that peak **3** represented a polyhydroxy daphnane type compound, which lacks the orthoester moiety at the C-ring. Thus, peaks **3** and **16** were identified as wikstroelide M (**3**) [14] and wikstrotoxin B (**16**) [11], respectively.

In the product ion spectra of peaks **6**, **9**, **10**, **14**, and **19**, the product ion generated by the loss of the $C_3H_4O_2$ unit was consistently observed at *m*/*z* 269 ($C_{17}H_{17}O_3$), indicating that these peaks corresponded to orthoester daphnanes with a substituent attached to C-12 (Figures 3B and S24). The product ions corresponding to substituents observed in these peaks were assignable as follows: a cinnamoyl moiety at *m*/*z* 131 (C_9H_7O), a coumaroyl moiety at *m*/*z* 147 ($C_9H_7O_2$), a feruloyl moiety at *m*/*z* 177 ($C_{10}H_9O_3$), a 2*E*,4*E*-decadienoyl moiety at *m*/*z* 151 ($C_{10}H_{15}O$), 95 (C_6H_7O), and 81 (C_5H_5O), a 2*E*,4*E*,6*E*-decatrienoyl moiety at *m*/*z* 149 ($C_{10}H_{13}O$) and 107 (C_9H_9O), and a 2*E*,4*E*-dodecadienoyl moiety at *m*/*z* 179 ($C_{12}H_{19}O$), 95 (C_6H_7O), and 81 (C_5H_5O). Namely, the feruloyl and 2*E*,4*E*,6*E*-decatrienoyl moieties were present in peak **6**, the coumaroyl and 2*E*,4*E*-decadienoyl moieties in peak **9**, the feruloyl and 2*E*,4*E*-decadienoyl moieties in peak **10**, and the cinnamoyl and 2*E*,4*E*,6*E*-decatrienoyl moieties in peak **14**. In peak **19**, only the product ions due to the 2*E*,4*E*-dodecadienoyl moiety were observed, but the molecular formula and the observation of product ions derived from the neutral loss of $C_2H_4O_2$ suggested the presence of the acetyl moiety. Thus, peaks **6**, **9**, **10**, **14**, and **19** were identified as acutilobin C (**6**) [36], daphneodorin D (**9**) [29], acutilobin D (**10**) [36], 12-*O*-(*E*)-cinnamoyl-9,13,14-ortho- (2*E*,4*E*,6*E*)-decatrienylidyne-5β,12β-dihydroxyresiniferonol-6α,7α-oxide (**14**) [40], and wikstroelide A (**19**) [13].

Peaks **1**, **2**, **4**, **5**, **7**, **8**, **11**, and **13** were identified as MDOs by their characteristic MS/MS fragmentation patterns. Although the number of oxygen functional group varied among these compounds, they were all characterized by the abundance of C_{30} to C_{28} product ions observed in the range of *m*/*z* 400–550. The molecular formula of peak **7** indicated the absence of acyl groups. In the product ion spectrum, the neutral loss associated with the macrocyclic ring was assigned to be $C_{10}H_{16}O$ as in compounds **18** and **20**. In addition, a series of C_{20} to C_{18} product ions were observed with successive losses of H_2O and CO from *m*/*z* 327 ($C_{20}H_{23}O_4$) (Figures 4A and S25). Peak **7** was suggested to possess the cyclopentanone A-ring structure based on the degree of unsaturation and was identified as pimelea factor S_6 (**7**) [37]. Peak **5** had a molecular formula that was 16 Da (OH) larger than peaks **18** and **20**. The product ion spectrum of peak **5** exhibited a series of C_{20} to C_{18} product ions below *m*/*z* 350, as observed in **18** and **20** (Figures 4B and S25). However, the neutral loss associated with the macrocyclic ring differed from **18** and **20**, where it was $C_{10}H_{16}O$ rather than $C_{10}H_{14}O$ in **5**. These observations indicated that **5** possesses an additional hydroxyl group at C-2′ of the macrocyclic ring compared to **18** and **20**, and was further identified as kraussianin (**5**) [35].

The pair of peaks **1** and **2**, as well as the pair of peaks **8** and **11**, had the same molecular formula and exhibited similar product ion spectra, indicating that each pair, like compounds **18** and **20**, was in a regioisomeric relationship. The product ion spectra of peaks **1** and **2** revealed three molecules of $C_7H_6O_2$ neutral loss originating from benzoyl acids and two molecules of $C_2H_4O_2$ neutral loss originating from acetic acids, as well as the neutral loss of $C_{10}H_{10}O$ associated with the macrocyclic ring. In addition, a series of C_{30} to C_{28} product ions were observed with successive losses of H_2O and CO from *m*/*z* 507 ($C_{30}H_{35}O_7$) (Figures 4C and S26). These observations suggested that peaks **1** and **2** correspond to daphneodorin B (**1**) or daphneodorin C (**2**) with the same molecular formula and combinations of acyl groups. By comparison of retention times, peaks **1** and **2** were identified as daphneodorin C (**1**) and daphneodorin B (**2**), respectively [29]. The product ion spectra of peaks **8** and **11** revealed the elimination of two molecules of $C_7H_6O_2$, indicating the presence of two benzoyloxy moieties (Figures 4C and S26). Additionally, the neutral loss associated with the macrocyclic ring was assigned to be $C_{10}H_{14}O$ as in **5** and the product ion pattern below m/z 350 was the same as **12**, which were identified as the regioisomer, stelleralide H (**8**) [38] and gnidimacrin (**11**) [39], respectively.

Peak **4** had a molecular weight 14 Da greater than peak **7** but the neutral loss associated with the macrocyclic ring was assigned to be $C_{10}H_{16}O$ as in peak **7**, suggesting that the daphnane skeleton was different from peak **7** (Figures 4D and S27). Based on the molecular formula and degree of unsaturation, peak **4** was identified as pimelotide C [34], which has the bicyclo[2.2.1]heptane A-ring structure. In the product ion spectrum of peak **13**, the observation of a loss of $C_7H_6O_2$ suggested the presence of a benzoyloxy moiety (Figures 4E and S27). Furthermore, the product ions observed within the range of m/z 400–550 were 2 Da smaller than those of peak **4**. Thus, peak **13** was identified as stelleralide C [5], which shared the bicyclo[2.2.1]heptane A-ring structure of peak **4** and a benzoyloxy moiety attached to C-18.

3. Materials and Methods

3.1. General Experimental Procedures

The NMR spectra were collected on a JEOL ECA-500 spectrometer (JEOL Ltd., Tokyo, Japan) with the deuterated solvent used as the internal reference. The ^{1}H-NMR spectra were performed at 500 MHz, and the ^{13}C-NMR spectra were generated at 125 MHz. HRESIMS was conducted using a Q-Exactive Hybrid Quadrupole Orbitrap mass spectrometer (Thermo Scientific, Waltham, MA, USA). The following columns were utilized for column chromatography: Diaion HP-20 (Mitsubishi Chemical Corporation, Tokyo, Japan), ODS (Chromatorex DM1020T, Fuji Silysia Chemical Ltd., Aichi, Japan) and silica gel (Chromatorex PEI MB 100-40/75, Fuji Silysia Chemical Ltd., Aichi, Japan) columns. For gradient HPLC, two JASCO/PU-2080 Plus Intelligent HPLC pumps (JASCO Corporation, Tokyo, Japan), equipped with an MX-2080-32 dynamic mixer (JASCO Corporation, Tokyo, Japan), a JASCO UV-970 Intelligent UV/vis detector, and an SSC-6800 fraction collector (JASCO Corporation, Tokyo, Japan), were utilized. For preparative HPLC, a Waters 515 HPLC pump (Waters Corporation, Massachusetts, USA), equipped with an ERC RefractoMax520 differential refractometer detector (Thermo Scientific, Waltham, MA, USA) and a Shimadzu SPD-10A UV-vis detector (Shimadzu, Kyoto, Japan), was utilized. For normal-phase HPLC separations, a silica gel column (YMC-Pack SIL, 5 μm, 250 × 20 mm) was utilized with a flow rate of 5.0 mL/min. For reversed-phase HPLC separations, an RP-C_{18} silica gel column (YMC-Actus Triart C_{18}, 5 μm, 150 × 20 mm) was utilized, with a flow rate of 8.0 mL/min.

3.2. Plant Material

The stems of *W. indica* were collected at Guangxi Province, People's Republic of China in February 2018 and identified by Dong Liang (Kunming Plant Classification and Biotechnology Co., Ltd., Kunming, China). A voucher specimen (accession number: 20201021) had been deposited in the herbarium of Shenyang Pharmaceutical University.

3.3. Extraction and Isolation

The air-dried whole plants of *W. indica* (1000 g) were cut into small pieces and extracted with 95% EtOH at room temperature to give an EtOH extract and a residue. The EtOH extract was concentrated (63.0 g), suspended in H_2O, and then partitioned with EtOAc. The EtOAc fraction (30.0 g) was subjected to Diaion HP-20 column chromatography, eluted with a stepwise gradient of MeOH/H_2O (from 5:5 to 10:0, v/v) to afford three fractions (E1 to E3). The E3 fraction (10.8 g) was subjected to ODS column chromatography, eluted with a stepwise gradient of MeOH–H_2O (from 7:3 to 10:0, v/v) to afford four subfractions (E3-1 to E3-4). Subfraction E3-2 (1840.3 mg) was subjected to silica gel column chromatography, eluted with a gradient of *n*-hexane–EtOAc–MeOH–HCOOH, to afford four subfractions (E3-2-1 to E3-2-5). Subfraction E3-2-3 (193.4 mg) was purified by RP-HPLC (70% CH_3CN, 80% CH_3CN) to give **12** (0.6 mg). Subfraction E3-2-2 (73.0 mg) was purified by RP-HPLC (85% CH_3CN) to give six subfractions (E3-2-2-1 to E3-2-2-6). Subfraction E3-2-2-3 (16.6 mg) was purified by RP-HPLC (80% CH_3CN), followed by NP-HPLC (*n*-hexane/AcOEt, 3:7) to give **15** (7.6 mg), **17** (1.6 mg), and **18** (1.3 mg). Subfraction E3-2-2-4 (4.4 mg) was purified by RP-HPLC (85% CH_3CN) to give **20** (1.9 mg) and **21** (1.0 mg).

3.4. LC-MS/MS Conditions

The LC-MS/MS analysis was performed using the same instruments and column as in previous experiments [20]. For LC conditions, the mobile phase comprised eluent A (distilled water with 0.1% formic acid) and B (acetonitrile with 0.1% formic acid), programmed as follows: 0–15 min, a linear gradient from 50% to 100% B, 15–18 min, 100% B, followed by column re-equilibration at 50% B for 10 min before the subsequent injection. For MS conditions, the in-source CID was set at 0 eV, and the resolution was 70,000 for full MS and 35,000 for full MS/data dependent (dd)-MS/MS modes. The AGC was established at 1E6 for full MS and 1E5 for dd-MS/MS. Data-dependent scanning was performed using HCD with the normalized collision energy at 15 eV. The extracted ion spectra were generated by extracting the following base peaks of each compounds within ± 5 ppm mass tolerance: m/z 1028.4283 $[M + NH_4]^+$ (**1**), m/z 1028.4264 $[M + NH_4]^+$ (**2**), m/z 585 $[M + H–H_2O]^+$ (**3**), m/z 547.2888 $[M + H]^+$ (**4**), m/z 655.3469 $[M + H]^+$ (**5**), m/z 719.3062 $[M + H]^+$ (**6**), m/z 533.3104 $[M + H]^+$ (**7**), m/z 775.3690 $[M + H]^+$ (**8**), m/z 691.3133 $[M + H]^+$ (**9**), m/z 721.3203 $[M + H]^+$ (**10**), m/z 775.3687 $[M + H]^+$ (**11**), m/z 912.4147 $[M + NH_4]^+$ (**12**), m/z 667.3107 $[M + H]^+$ (**13**), m/z 673.2999 $[M + H]^+$ (**14**), m/z 533.3107 $[M + H]^+$ (**15**), m/z 557.3109 $[M + H]^+$ (**16**), m/z 675.3151 $[M + H]^+$ (**17**), m/z 639.3527 $[M + H]^+$ (**18**), m/z 643.3458 $[M + H]^+$ (**19**), m/z 639.3525 $[M + H]^+$ (**20**), and m/z 585.3409 $[M + H]^+$ (**21**). All data collected in the profile mode were acquired and processed using Thermo Xcalibur 4.1 software.

4. Conclusions

This study represents the first comprehensive identification of 21 daphnane diterpenoids from the stems of *W. indica* through a combination of LC-MS guided isolation and MS/MS fragmentation elucidation. The investigation revealed that *W. indica* contained structurally diverse daphnane diterpenoids, including orthoester daphnane type, polyhydroxy daphnane type, and macrocyclic daphnane orthoester type compounds. The application of MS/MS fragmentation elucidation for structural analysis enabled the rapid and precise identification of these diterpenoids within crude plant extracts. This methodology holds great promise for future research endeavors aimed at discovering bioactive diterpenoids from plants of the Thymelaeaceae family.

Supplementary Materials: The following supporting information can be downloaded at: https://www.mdpi.com/article/10.3390/plants12203620/s1, Figure S1: Total ion chromatograms and extracted ion chromatograms in the positive ion mode from the (A) E3-2 fraction, (B) E3-3 fraction, and (C) E3-4 fraction; Tables S1 and S2: ^1H- (500 MHz) and ^{13}C- (125 MHz) NMR data of compounds **12**, **15**, **17**, **18**, **20**, and **21** ($CDCl_3$); Figures S2–S21: NMR spectra and HRESIMS data of compounds **12**, **15**, **17**, **18**, **20**, and **21**; Figures S22–S27: Product ion spectra of peaks **1–11** and **13–21** obtained from the protonated molecular ion peak in positive ion mode, and **12** obtained from the ammonium adduct ion peak positive ion mode.

Author Contributions: Conceptualization, W.L.; methodology, M.Z., K.O. and W.L.; formal analysis, M.Z. and K.O.; investigation, M.Z., R.T. and K.O.; resources, D.Z. and N.L.; writing—original draft preparation, M.Z. and K.O.; writing—review and editing, K.O., T.K., N.L. and W.L.; supervision, K.O., T.K. and W.L.; project administration, K.O. and W.L.; funding acquisition, M.Z. and W.L. All authors critically reviewed and revised the manuscript draft and approved the final version for submission. All authors have read and agreed to the published version of the manuscript.

Funding: The investigation was supported by the Japan Society for the Promotion of Science KAKENHI 21K06619 (W.L.). This work was also supported by the Sasakawa Scientific Research Grant from The Japan Science Society (No. 2022-3015) (M.Z.).

Data Availability Statement: All new research data were presented in this contribution.

Acknowledgments: All authors express their appreciation to the editorial board and reviewer for their suggestions and efforts in reviewing this paper. The authors also appreciate the editor for his assistance during the review process.

Conflicts of Interest: The authors declare no competing financial interest.

References

1. Tsugawa, H.; Rai, A.; Saito, K.; Nakabayashi, R. Metabolomics and Complementary Techniques to Investigate the Plant Phytochemical Cosmos. *Nat. Prod. Rep.* **2021**, *38*, 1729–1759. [CrossRef]
2. Azmir, J.; Zaidul, I.S.M.; Rahman, M.M.; Sharif, K.M.; Mohamed, A.; Sahena, F.; Jahurul, M.H.A.; Ghafoor, K.; Norulaini, N.A.N.; Omar, A.K.M. Techniques for Extraction of Bioactive Compounds from Plant Materials: A Review. *J. Food Eng.* **2013**, *117*, 426–436. [CrossRef]
3. Wolfender, J.-L.; Nuzillard, J.-M.; van der Hooft, J.J.J.; Renault, J.-H.; Bertrand, S. Accelerating Metabolite Identification in Natural Product Research: Toward an Ideal Combination of Liquid Chromatography-High-Resolution Tandem Mass Spectrometry and NMR Profiling, in silico Databases, and Chemometrics. *Anal. Chem.* **2019**, *91*, 704–742. [CrossRef]
4. Kupchan, S.M.; Baxter, R.L. Mezerein: Antileukemic Principle Isolated from *Daphne mezereum* L. *Science* **1975**, *187*, 652–653. [CrossRef] [PubMed]
5. Asada, Y.; Sukemori, A.; Watanabe, T.; Malla, K.J.; Yoshikawa, T.; Li, W.; Koike, K.; Chen, C.H.; Akiyama, T.; Qian, K.; et al. Stelleralides A–C, Novel Potent Anti-HIV Daphnane-Type Diterpenoids from *Stellera chamaejasme* L. *Org. Lett.* **2011**, *13*, 2904–2907. [CrossRef] [PubMed]
6. Appendino, G.; Szallasi, A. Euphorbium: Modern Research on its Active Principle, Resiniferatoxin, Revives an Ancient Medicine. *Life Sci.* **1997**, *60*, 681–696. [CrossRef] [PubMed]
7. Bailly, C. Yuanhuacin and Related Anti-Inflammatory and Anticancer Daphnane Diterpenes from Genkwa Flos-An Overview. *Biomolecules* **2022**, *12*, 192. [CrossRef]
8. He, W.; Cik, M.; Lesage, A.; Van der Linden, I.; De Kimpe, N.; Appendino, G.; Bracke, J.; Mathenge, S.G.; Mudida, F.P.; Leysen, J.E.; et al. Kirkinine, a New Daphnane Orthoester with Potent Neurotrophic Activity from *Synaptolepis kirkii*. *J. Nat. Prod.* **2000**, *63*, 1185–1187. [CrossRef]
9. He, W.; Cik, M.; Van Puyvelde, L.; Van Dun, J.; Appendino, G.; Lesage, A.; Van der Lindin, I.; Leysen, J.E.; Wouters, W.; Mathenge, S.G.; et al. Neurotrophic and Antileukemic Daphnane Diterpenoids from *Synaptolepis kirkii*. *Bioorg. Med. Chem.* **2002**, *10*, 3245–3255. [CrossRef]
10. Otsuki, K.; Li, W. Tigliane and Daphnane Diterpenoids from Thymelaeaceae family: Chemistry, Biological Activity, and Potential in Drug Discovery. *J. Nat. Med.* **2023**, *77*, 625–643. [CrossRef]
11. Jolad, S.D.; Hoffmann, J.J.; Timmermann, B.N.; Schram, K.H.; Cole, J.R.; Bates, R.B.; Klenck, R.E.; Tempesta, M.S. Daphnane Diterpenes from *Wikstroemia monticola*: Wikstrotoxins A–D, Huratoxin, and Excoecariatoxin. *J. Nat. Prod.* **1983**, *46*, 675–680. [CrossRef]
12. Dagang, W.; Sorg, B.; Adolf, W.; Seip, E.H.; Hecker, E. Oligo- and Macrocyclic Diterpenes in Thymelaeaceae and Euphorbiaceae Occurring and Utilized in Yunnan (Southwest China) 3. Two New Daphnane type 9,13,14-Orthoesters from *Wikstroemia mekongenia*. *Phytother. Res.* **1993**, *7*, 72–75. [CrossRef]
13. Abe, F.; Iwase, Y.; Yamauchi, T.; Kinjo, K.; Yaga, S. Daphnane Diterpenoids from the Bark of *Wikstroemia retusa*. *Phytochemistry* **1997**, *44*, 643–647. [CrossRef]
14. Abe, F.; Iwase, Y.; Yamauchi, T.; Kinjo, K.; Yaga, S.; Ishii, M.; Iwahana, M. Minor Daphnane-type Diterpenoids from *Wikstroemia retusa*. *Phytochemistry* **1998**, *47*, 833–837. [CrossRef] [PubMed]
15. Khong, A.; Forestieri, R.; Williams, D.E.; Patrick, B.O.; Olmstead, A.; Svinti, V.; Schaeffer, E.; Jean, F.; Roberge, M.; Andersen, R.J.; et al. A Daphnane Diterpenoid Isolated from *Wikstroemia polyantha* Induces an Inflammatory Response and Modulates miRNA Activity. *PLoS ONE* **2012**, *7*, e39621. [CrossRef]
16. Guo, J.; Zhang, J.; Shu, P.; Kong, L.; Hao, X.; Xue, Y.; Luo, Z.; Li, Y.; Li, G.; Yao, G.; et al. Two new diterpenoids from the buds of *Wikstroemia chamaedaphne*. *Molecules* **2012**, *17*, 6424–6433. [CrossRef]
17. Guo, J.; Tian, J.; Yao, G.; Zhu, H.; Xue, Y.; Luo, Z.; Zhang, J.; Zhang, Y.; Zhang, Y. Three New 1α-alkyldaphnane-type Diterpenoids from the Flower Buds of *Wikstroemia chamaedaphne*. *Fitoterapia* **2015**, *106*, 242–246. [CrossRef]
18. Li, S.F.; Jiao, Y.Y.; Zhang, Z.Q.; Chao, J.B.; Jia, J.; Shi, X.L.; Zhang, L.W. Diterpenes from buds of *Wikstroemia chamaedaphne* showing anti-hepatitis B virus activities. *Phytochemistry* **2018**, *151*, 17–25. [CrossRef]
19. Liu, Y.Y.; Liu, Y.P.; Wang, X.P.; Qiao, Z.H.; Yu, X.M.; Zhu, Y.Z.; Xie, L.; Qiang, L.; Fu, Y.H. Bioactive Daphnane Diterpenes from *Wikstroemia chuii* with their Potential Anti-Inflammatory Effects and Anti-HIV Activities. *Bioorg. Chem.* **2020**, *105*, 104388. [CrossRef]
20. Otsuki, K.; Zhang, M.; Kikuchi, T.; Tsuji, M.; Tejima, M.; Bai, Z.S.; Zhou, D.; Huang, L.; Chen, C.H.; Lee, K.H.; et al. Identification of Anti-HIV Macrocyclic Daphnane Orthoesters from *Wikstroemia ligustrina* by LC-MS Analysis and Phytochemical Investigation. *J. Nat. Med.* **2021**, *75*, 1058–1066. [CrossRef]
21. Shirai, K. *Tyuyakudaiziten*; Shanghai Scientific & Technical Publishers, Shougakukan, Inc.: Tokyo, Japan, 1985; pp. 2701–2703.

22. Jegal, J.; Park, N.J.; Lee, S.Y.; Jo, B.G.; Bong, S.K.; Kim, S.N.; Yang, M.H. Quercitrin, the Main Compound in *Wikstroemia indica*, Mitigates Skin Lesions in a Mouse Model of 2,4-Dinitrochlorobenzene-Induced Contact Hypersensitivity. *Evid.-Based Complement. Alternat. Med.* **2020**, *2020*, 4307161. [CrossRef]
23. Lee, S.Y.; Park, N.J.; Jegal, J.; Jo, B.G.; Choi, S.; Lee, S.W.; Uddin, M.S.; Kim, S.N.; Yang, M.H. Suppression of DNCB-Induced Atopic Skin Lesions in Mice by *Wikstroemia indica* Extract. *Nutrients* **2020**, *12*, 173. [CrossRef] [PubMed]
24. Huang, Y.C.; Huang, C.P.; Lin, C.P.; Yang, K.C.; Lei, Y.J.; Wang, H.P.; Kuo, Y.H.; Chen, Y.J. Naturally Occurring Bicoumarin Compound Daphnoretin Inhibits Growth and Induces Megakaryocytic Differentiation in Human Chronic Myeloid Leukemia Cells. *Cells* **2022**, *11*, 3252. [CrossRef] [PubMed]
25. Chen, Y.; Fu, W.W.; Sun, L.X.; Wang, Q.; Qi, W.; Yu, H. A New Coumarin from *Wikstroemia indica* (L.) C. A. Mey. *Chin. Chem. Lett.* **2009**, *20*, 592–594. [CrossRef]
26. Bailly, C. Bioactive Biflavonoids from *Wikstroemia indica* (L.) C.A. Mey. (Thymelaeaceae): A Review. *Trends Phytochem. Res.* **2021**, *5*, 190–198.
27. Wang, L.-Y.; Unehara, N.; Kitanaka, S. Lignans from the Roots of *Wikstroemia indica* and Their DPPH Radical Scavenging and Nitric Oxide Inhibitory Activities. *Chem. Pharm. Bull.* **2005**, *53*, 1348–1351. [CrossRef]
28. Wang, L.-Y.; Unehara, T.; Kitanaka, S. Anti-Inflammatory Activity of New Guaiane Type Sesquiterpene from *Wikstroemia indica*. *Chem. Pharm. Bull.* **2005**, *53*, 137–139. [CrossRef]
29. Otsuki, K.; Li, W.; Asada, Y.; Chen, C.H.; Lee, K.H.; Koike, K. Daphneodorins A–C, Anti-HIV Gnidimacrin Related Macrocyclic Daphnane Orthoesters from *Daphne odora*. *Org. Lett.* **2020**, *22*, 11–15. [CrossRef]
30. Tan, L.; Otsuki, K.; Zhang, M.; Kikuchi, T.; Okayasu, M.; Azumaya, I.; Zhou, D.; Li, N.; Huang, L.; Chen, C.-H.; et al. Daphnepedunins A–F, Anti-HIV Macrocyclic Daphnane Orthoester Diterpenoids from *Daphne pedunculata*. *J. Nat. Prod.* **2022**, *85*, 2856–2864. [CrossRef]
31. Zhao, H.D.; Lu, Y.; Yan, M.; Chen, C.H.; Morris-Natschke, S.L.; Lee, K.H.; Chen, D.F. Rapid Recognition and Targeted Isolation of Anti-HIV Daphnane Diterpenes from *Daphne genkwa* Guided by UPLC-MS^n. *J. Nat. Prod.* **2020**, *83*, 134–141. [CrossRef]
32. Trinel, M.; Jullian, V.; Le Lamer, A.C.; Mhamdi, I.; Mejia, K.; Castillo, D.; Cabanillas, B.J.; Fabre, N. Profiling of *Hura crepitans* L. latex by Ultra-High-Performance Liquid Chromatography/Atmospheric Pressure Chemical Ionisation Linear Ion Trap Orbitrap Mass Spectrometry. *Phytochem. Anal.* **2018**, *29*, 627–638. [CrossRef] [PubMed]
33. Zhang, M.; Otsuki, K.; Kato, S.; Ikuma, Y.; Kikuchi, T.; Li, N.; Koike, K.; Li, W. A feruloylated acylglycerol isolated from *Wikstroemia pilosa* and its Distribution in Ten plants of *Wikstroemia* species. *J. Nat. Med.* **2022**, *76*, 680–685. [CrossRef] [PubMed]
34. Hayes, P.Y.; Chow, S.; Somerville, M.J.; Fletcher, M.T.; De Voss, J.J. Daphnane- and Tigliane-type Diterpenoid Esters and Orthoesters from *Pimelea elongata*. *J. Nat. Prod.* **2010**, *73*, 1907–1913. [CrossRef] [PubMed]
35. Borris, R.P.; Cordell, G.A. Studies of the Thymelaeaceae II. Antineoplastic principles of *Gnidia kraussiana*. *J. Nat. Prod.* **1984**, *47*, 270–278. [CrossRef] [PubMed]
36. Huang, S.Z.; Zhang, X.J.; Li, X.Y.; Kong, L.M.; Jiang, H.Z.; Ma, Q.Y.; Liu, Y.Q.; Hu, J.M.; Zheng, Y.T.; Li, Y.; et al. Daphnane-type diterpene esters with cytotoxic and anti-HIV-1 activities from *Daphne acutiloba* Rehd. *Phytochemistry* **2012**, *75*, 99–107. [CrossRef]
37. Hayes, P.Y.; Chow, S.; Somerville, M.J.; De Voss, J.J.; Fletcher, M.T. Pimelotides A and B, Diterpenoid Ketal-lactone Orthoesters with An Unprecedented Skeleton from *Pimelea elongata*. *J. Nat. Prod.* **2009**, *72*, 2081–2083. [CrossRef]
38. Yan, M.; Lu, Y.; Chen, C.H.; Zhao, Y.; Lee, K.H.; Chen, D.F. Stelleralides D–J and Anti-HIV Daphnane Diterpenes from *Stellera chamaejasme*. *J. Nat. Prod.* **2015**, *78*, 2712–2718. [CrossRef]
39. Kupchan, S.M.; Shizuri, Y.; Murae, T.; Swenny, J.G.; Haynes, H.R.; Shen, M.S.; Barrick, J.C.; Bryan, A.F.; vander Helm, D.; Wu, K.K. Letter: Gnidimacrin and Gnidimacrin 20-palmitate, Novel Macrocyclic Antileukemic Diterpenoid Esters from *Gnidia subcordata*1,2. *J. Am. Chem. Soc.* **1976**, *98*, 5719–5720. [CrossRef]
40. Ohigashi, H.; Hirota, M.; Ohtsuka, T.; Koshimizu, K.; Fujiki, H.; Suganuma, M.; Yamaizumi, Z.; Sugimura, T. Resiniferonol-Related Diterpene Esters from *Daphne odora* Thunb. and Their Ornithine Decarboxylase-Inducing Activity in Mouse Skin. *Agric. Biol. Chem.* **1982**, *46*, 2605–2608. [CrossRef]
41. Pettit, G.R.; Zou, J.C.; Goswami, A.; Cragg, G.M.; Schmidt, J.M. Antineoplastic agents, 88. *Pimelea prostrata*. *J. Nat. Prod.* **1983**, *46*, 563–568. [CrossRef]
42. Sakata, K.; Kawazu, K.; Mitsui, T. Studies on a Piscicidal Constituent of *Hura crepitans*. *Agric. Biol. Chem.* **1971**, *35*, 1084–1091.

plants

MDPI

Communication

Argentatin Content in Guayule Leaves (*Parthenium argentatum* A. Gray)

María Mercedes García-Martínez [1,2], Beatriz Gallego [3], Guayente Latorre [2], María Engracia Carrión [4], Miguel Ángel De la Cruz-Morcillo [4], Amaya Zalacain [2] and Manuel Carmona [4,*]

1 Instituto Técnico Agronómico Provincial de Albacete (ITAP), Parque empresarial Campollano, 2ª Avenida, 61, 02007 Albacete, Spain; mariamercedes.garcia@uclm.es
2 E.T.S.I. Agronómica, de Montes y Biotecnología (ETSIAMB), Cátedra de Química Agrícola, Universidad de Castilla-La Mancha, Avda. de España s/n, 02071 Albacete, Spain; guayente.latorre@uclm.es (G.L.); amaya.zalacain@uclm.es (A.Z.)
3 Instituto de Toxicología de la Defensa, Hospital Central de la Defensa Gómez Ulla, Gta Ejército 1, 28047 Madrid, Spain; bgaligl@mde.es
4 Institute for Regional Development (IDR), Food Quality Research Group, Universidad de Castilla-La Mancha Campus Universitario s/n, 02071 Albacete, Spain; mengracia.carrion@uclm.es (M.E.C.); miguelangel.cruz@uclm.es (M.Á.D.l.C.-M.)
* Correspondence: manuel.carmona@uclm.es

Abstract: Approximately one-third of the waste biomass from the cultivation of guayule (*Parthenium argentatum* A. Gray) for natural rubber production is leaf tissue; however, whether it can be valorized is not known. Guayulins and argentatins are potential high-value products that can be recovered from guayule resin during rubber/latex processing. Argentatins are highly abundant in guayule stem resin; however, unlike the guayulins, their occurrence in leaves has not been investigated. The present study determined the content of argentatins and isoargentatins A and B in the leaves of a pure guayule accession (R1040) and two hybrids (CAL-1 and AZ-2) under conditions of irrigation and non-irrigation. The resin content in leaves was ~10%, which provides a suitable starting point for economic exploitation. The highest production of argentatins occurred in plants under irrigation, with yields of 4.2 and 3.6 kg ha^{-1} for R1040 and AZ-2, respectively. The R1040 accession had the highest percentage of resin and the greatest total argentatin content (24.5 g kg^{-1} dried leaf), principally due to the abundance of argentatin A. Contrastingly, CAL-1 consistently showed the lowest argentatin content based on dried leaf weight and production (0.6 kg ha^{-1}). The substantial abundance of argentatins in guayule leaves suggests the potential for future exploitation.

Keywords: guayule leaves; argentatin A; isoargentatin A; argentatin B; isoargentatin B

1. Introduction

Guayule (*Parthenium argentatum* A. Gray) is the best candidate for countries consuming large quantities of rubber with arid and semi-arid agricultural areas where it is not possible to cultivate *Hevea brasiliensis* [1]. It was used as a source of rubber in the past, especially for the manufacturing of tyres; however, it is now estimated that it could only become an economically competitive crop again if all its co-products were exploited, rather than just rubber [2–4].

Approximately 30% of the waste biomass from the cultivation of guayule is leaf tissue [5], which is typically discarded prior to bulk rubber extraction [2]. This process occurs because leaf materials can reduce rubber recovery due to the formation of very fine particles that clog the filtration systems and increase the ash content associated with rubber degradation and quality reduction [2,6]. The complete morphological characterization of guayule leaves was performed within 27 guayule accessions in Spain, grouping them into three main groups depending on their morphological parameters, which, in general, were apiculate leaves with the petiole delimited based on leaf margin, serrate or lanceolate types [7]. Differences identified within the recent USDA germplasm collection confirm that growth environment has a major influence on leaf morphological characterization [8].

Other studies proposed the valorization of different plants from their leaves [9–12]. Early studies on the valorization of guayule by-products focused on the extraction of essential and volatile oils from stem resin [13,14], while more recent attention was paid to phenolic compounds [7,15,16], particularly hydroxycinnamic and hydroxybenzoic acids, flavones, flavanols and anthraquinones [7,15]. In addition to the stem resin, a significant proportion of total guayulin content is found in leaf tissue, where the four major guayulins (A–D) are observed in a more consistent relationship than in the stem [17], whereas rubber content in the leaves is discarded [18].

Beyond the commercialization of the guayulins, a large number of other extractable secondary metabolites that could be economically exploited are present in relevant amounts in guayule stem resin, including triterpenes and their derivatives, known as the argentatins. Recently, Gallego et al. [19] identified, for first time, the argentatins and derivates profile in 27 different guayule accessions with LC-CR-MS technology (Orbitrap®). There is limited information on argentatin accumulation as their routine quantification is challenging because the structure lacks double bonds and shows a weak chromophoric functionality; however, Gallego et al. [20] recently proposed their quantification using a refractive index detector after their chromatographic separation. The most abundant argentatins in guayule resin correspond to argentatin A ($C_{30}H_{48}O_4$) and B ($C_{30}H_{48}O_3$), as representative of the cycloartenol-type triterpenes, and isoargentatin A ($C_{30}H_{48}O_4$) and isoargentatin B ($C_{30}H_{48}O_3$), with lanostane-type triterpenes (Figure 1).

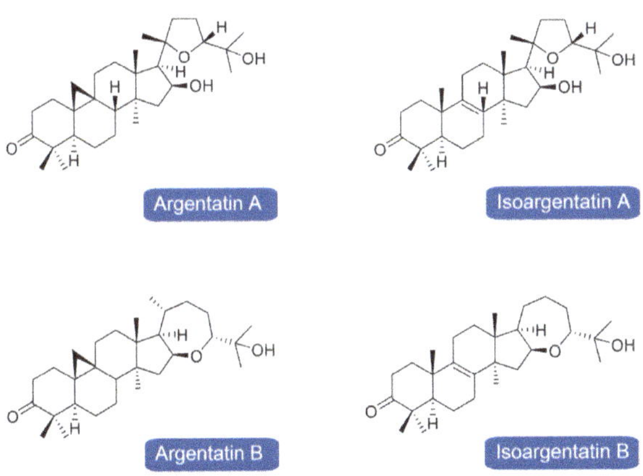

Figure 1. Structure of most abundant argentatins and isoargentatins in guayule accessions [19].

Argentatins attracted considerable attention, in particular argentatins A and B, because they have several biological activities that make them potentially useful as pharmacophores. Argentatin A has antimicrobial activity against pathogenic bacteria [21], and argentatins A, B and D have anti-inflammatory activity [22]. In vitro cytotoxic activity against a broad panel of human cancer cell lines was previously attributed to argentatin A and B [22–25], and in vivo in a mouse xenograft model of human colon cancer was previously associated with argentatin A [25]. More recent studies explored the use of argentatins as precursors for the synthesis of compounds with enhanced activity and lower toxicity [26–28].

The production of argentatins in guayule stems depends on the accession used and the time of harvest, with the spring and summer periods showing maximum accumulation [20]. To the best of our knowledge, no previous study examined whether argentatins are produced in guayule leaves. Therefore, the aim of the present study was to explore the content of the four major argentatins (argentatins and isoargentatins A and B) in the leaves of three guayule and hybrid accessions under different agronomic conditions (irrigation and non-irrigation). The content of argentatins in leaves was then compared with the content in stems in the same accessions.

2. Results

2.1. Leaf Resin Extraction and Quantitation of Argentatins

The leaf resin content from the three guayule accessions ranged from 9.5% to 11.5% (Table 1), and the standard deviation range for all accessions was very small, with the exception of AZ-2 under non-irrigation conditions. No significant differences in leaf resin content were found between the three accessions under irrigation conditions (Table 1); however, the resin content of R1040 was significantly greater under non-irrigation than irrigation conditions and was also significantly greater than that of the other two accessions.

Table 1. Descriptive statistics for argentatins and isoargentatins A and B content in leaves under different conditions of irrigation.

Crop System	Accession	Statistical Parameters	Resin Yield (%)	Samples Analyzed via RID (g kg^{-1} Dried Leaf)				
				aA	isoaA	aB	isoaB	aT
Irrigated	AZ-2	Min.	9.14	6.86	2.52	6.33	0.67	
		Max.	9.96	6.98	2.57	6.55	0.69	
		Mean	9.55 A, a	6.92 A, a, δ	2.54 A, b, β	6.44 A, b, γ	0.68 B, b, α	16.58 A, a
		SD	0.58	0.08	0.04	0.16	0.01	
	CAL-1	Min.	9.18	6.02	1.25	6.78	0.54	
		Max.	9.85	6.47	1.35	7.33	0.57	
		Mean	9.52 A, a	6.25 B, a, γ	1.30 A, a, β	7.05 A, b, δ	0.56 A, a, α	15.16 B, a
		SD	0.48	0.32	0.07	0.38	0.02	
	R1040	Min.	9.44	9.50	2.63	5.44	0.64	
		Max.	9.97	9.91	2.82	5.64	0.67	
		Mean	9.71 A, a	9.70 A, b, δ	2.73 A, b, β	5.54 A, a, γ	0.65 A, b, α	18.62 A, b
		SD	0.38	0.29	0.13	0.14	0.02	
Non-irrigated	AZ-2	Min.	8.28	7.50	1.76	5.37	0.38	
		Max.	10.79	9.74	2.25	7.04	0.47	
		Mean	9.53 A, a	8.62 A, b, δ	2.00 A, a, β	6.21 A, a, γ	0.43 A, a, α	17.26 A, b
		SD	1.77	1.59	0.35	1.18	0.07	
	CAL-1	Min.	9.80	3.79	1.37	5.99	0.50	
		Max.	9.98	3.83	1.40	6.39	0.52	
		Mean	9.89 A, a	3.81 A, a, γ	1.38 A, a, β	6.19 A, a, δ	0.51 A, a, α	11.89 A, a
		SD	0.13	0.03	0.02	0.28	0.01	
	R1040	Min.	11.15	13.03	3.19	6.58	0.85	
		Max.	11.92	13.78	3.53	7.10	0.92	
		Mean	11.53 B, b	13.41 B, c, δ	3.36 B, b, β	6.84 A, a, γ	0.88 B, b, α	24.50 B, c
		SD	0.54	0.53	0.24	0.37	0.05	

Note: aA, argentatin A; isoaA, isoargentatin A; aB, argentatin B; isoaB, isoargentatin B; aT, total argentatins and isoargentatins A and B; RID, refractive index detection; SD, standard deviation. Different letters indicate significant differences between different crop system and accessions at 95% level of confidence. Uppercase letters refer to an ANOVA comparing differences between crop systems for same accessions; lowercase letters refer to an ANOVA comparing differences between accessions within same crop system; greek letters refer to an ANOVA comparing differences between argentatins within same accession.

In relation to total argentatin content (aT), the R1040 accession showed the highest total content relative to AZ-2 and CAL-1 (R1040 > AZ-2 = CAL-1) under both agronomic conditions, due to its significantly greater content of aA. Under non-irrigated conditions, aT content remained highest in R1040, and was even greater than under irrigated conditions. Contrastingly, the aB content was greater in AZ-2 and CAL-1 hybrids than in R1040 under irrigation conditions, whereas no significant differences were observed under non-irrigation conditions (Table 1).

2.2. Argentatin Production per Hectare

Overall, the leaf yield (per hectare) was higher with irrigation than without, with a mean yield of 215.6 kg ha^{-1} for irrigated plots of the three accessions and 74.9 kg ha^{-1} for non-irrigated plots (Table 2). When irrigated, R1040 produced a greater leaf yield per hectare (258.4 kg), but both AZ-2 and R1040 produced the highest total argentatin yield (kg ha^{-1}) when considering the content of argentatins (g kg^{-1} dried leaf). Likewise, AZ-2 and R1040 produced the highest content of aA and isoaA. No significant differences were observed in the production of aB between the three accessions.

Table 2. Leaf yield per accession and argentatin production per hectare.

Crop System	Accession	Leaf Yield per ha (kg)	Production per ha (kg)				Total Production per ha (kg)
			aA	isoaA	aB	isoaB	
Irrigated	AZ-2	207.19 B, a	1.50 B, ab	0.55 B, ab	1.39 B, a	0.15 B, a	3.59 B, ab
	CAL-1	181.20 B, a	1.07 B, a	0.22 B, a	1.21 B, a	0.10 B, a	2.60 B, a
	R1040	258.44 B, b	2.19 B, b	0.62 B, b	1.25 B, a	0.15 A, a	4.21 B, b
Non-irrigated	AZ-2	81.20 A, a	0.73 A, ab	0.17 A, ab	0.53 A, a	0.04 A, a	1.47 A, ab
	CAL-1	50.74 A, a	0.19 A, a	0.07 A, a	0.31 A, a	0.03 A, a	0.59 A, a
	R1040	92.72 A, a	1.35 A, b	0.34 A, b	0.69 A, a	0.09 A, b	2.47 A, b

Note: aA, argentatin A; isoaA, isoargentatin A; aB, argentatin B; isoaB, isoargentatin B; aT, total argentatins and isoargentatins A and B. Different letters between rows indicate significant differences between different crop system and accessions at 95% level of confidence. Uppercase letters refer to an ANOVA that compares differences between crop system for same accessions; lowercase letters refer to an ANOVA that compares differences between accessions within same crop system.

3. Discussion

The leaf resin content from the three guayule accessions studied is similar to the values reported for stem resin content by Gallego et al. [20], and is a suitable starting point for its industrial exploitation. Moreover, the resin percentage was consistent with values reported by Dehghanizadeh et al. [2] and Rozalén et al. [17], who obtained a yield of 9–10% in plants of the same age, and was much greater than the 4% reported in older plants (15–18 years old) by Spano et al. [29]. No significant differences in leaf resin content were found between the three accessions under irrigation conditions; however, R1040 resin content was clearly highest under non-irrigation conditions.

In relation to total argentatin content (aT), R1040 and AZ-2 accessions produced the highest content under both agronomic conditions. Analysis of all argentatins and isoargentatins revealed two distinct profiles: the content of aB was greater than aA in CAL-1, whereas the opposite was seen for AZ-2 and R1040. In contrast, the content of isoaA and isoaB was maintained for all three accessions (isoaA > isoaB) under both agronomic conditions.

In terms of leaf production per hectare, irrigation yields a higher content than non-irrigation. Through comparing the results of argentatins and isoargentatins obtained in leaves with their content in guayule stems [19], it is evident that the values obtained for the same accessions by Gallego and colleagues are much higher. Nonetheless, if the same accessions are compared with the stems of plants of the same age and under the same conditions [20], the contents of aA and aB are very similar.

The argentatin content in guayule leaves ranged between 1.2 and 2.7% dry weight, which is a credible concentration for a bioactive compound in a vegetable by-product with proven antitumor activity [25]. The semi-synthesis of argentatins A–C analogs was also successfully performed in an effort to increase their cytotoxicity [26]. Notably, the production of new drugs is largely (64%) derived from natural products, generally via semi-synthesis [30,31]. For example, podophyllotoxin is the starting material for the important antineoplastic drugs etoposide and teniposide. Podophyllotoxin is found in a concentration of ~1.3% in the rhizome of Podophyllum hexandrum [31,32]. Another example is paclitaxel, traded under the brand name Taxol, which is the most successful anticancer phyto-drug developed in the last 50 years. The content of the paclitaxel precursor in Taxus baccatta (European Yew) is even lower than that of podophyllotoxin, being around 0.01–0.03% [33]. It should be noted that the content of argentatins in leaves is much higher than other compounds that have been considered to valorize guayule leaves, including phenolic compounds (0.8%, as reported by Jara et al. [7] or their essential oil content, which were reported to be 1.04% [13].

4. Conclusions

A resin content of almost 10% in leaves is a relevant amount for its commercial exploitation, either directly for raw resin applications or for the isolation of compounds of interest, such as the argentatins. In the present study, the highest total argentatin production was achieved with R1040 and AZ-2 accessions under irrigation, together with the highest production of aA and isoaA. The production of aB and isoaB can be achieved indistinctly under irrigation conditions with the three tested accessions.

5. Materials and Methods

5.1. Guayule Samples

The pure R1040 guayule accession and the CAL-1 and AZ-2 hydrids were selected for study under irrigated and non-irrigated conditions. The experimental plot (0.3 ha) was planted in May 2019 in Santa Cruz de la Zarza, Toledo, Spain. A randomized complete block design with four replicates per accession was utilized for analysis. A sample of three adjacent plants was taken for each of the four replicates, giving a total of twelve plants from each accession. Plants (24 months from sowing) were manually harvested and packaged into kraft bags. Whole plants were dried for 48 h at 60 °C to achieve a moisture content of ~12%. Leaves and stems were then manually separated, and the dry biomass weight was calculated from each fraction. Leaves and stems were grounded using the following two-step procedure: in the first step, the material (30 g) was ground into 2-mm-sized particles using a hammer grinder (Mader 57075, Mealhada, Portugal) for 10–15 s; and in the second step, the material was ground into particles of 0.5 mm using a centrifuge grinder (Retsch ZM 1000, Haan, Germany) for 2 min. Dried ground samples were stored in screw cap-closed glass vessels at room temperature. The moisture content of the samples was measured using a halogen lamp moisture balance, model XM-120T (Cobos, Barcelona, Spain) at 105 °C. When moisture loss was less than 0.1% in 180 s, the material was considered to have reached a constant mass.

5.2. Resin Determination

Extraction of the resin fraction was performed in an ASE E-914 Speed Extractor (BUCHI, Postfach, Switzerland), as described by Rozalén et al. [34]. Resin extracts were collected in 240-mL flasks and transferred to pre-weighed flasks equilibrated in a desiccator for 30 min. Solvent evaporation was performed in a Multivapor BUCHI P-6 (Postfach, Switzerland) parallel system at 150 mbar and 50 °C. After evaporation, the pre-weighed flasks were stored for 60 min in a desiccator before final weighing. The resin percentage was determined gravimetrically through considering the dry weight. Each sample was twice extracted.

5.3. Argentatins Determination via HPLC-RID

Twenty microliters of resin dissolved in ethanol at 10 mg mL^{-1} and filtered (0.22 μm) was injected into an Agilent 1200 high-performance liquid chromatography (HPLC) system (Agilent Technologies, Palo Alto, CA, USA) equipped with a refractive index detector (RID) (Agilent, G1362A). Separation was performed on a reverse-phase ACE Excel 3 C18-PentaFluorPhenyl column (150 × 4.6 mm, 3-μm particle size) protected with an ACE Excel HPLC Pre-column Filter (0.5-μm particle size) (both from Advanced Chromatography Technologies Ltd., Reading, Berkshire, UK). An isocratic program was used with acetonitrile (60%) and Milli Q-grade water (40%) as solvents. The analysis time was 60 min. Agilent ChemStation software (version B.03.01) was used for quantification of argentatins and isoargentatins A and B (aA, isoaA, aB, isoaB), using the aA (95.9% purity) and aB (92.4% purity) standards kindly provided by Dr. Mariano Martínez-Vázquez (Universidad Nacional Autónoma de México). Quantification of argentatins using the RID was carried out from the corresponding six-level calibration curve: aA (10–1000 mg L^{-1}; r^2 = 0.9976) and aB (50–1000 mg L^{-1}; r^2 = 0.9980).

Author Contributions: Investigation, M.M.G.-M., B.G., G.L. and M.Á.D.l.C.-M.; Visualization, M.M.G.-M., A.Z. and M.C.; Writing—Original Draft Preparation, M.M.G.-M., A.Z. and M.C.; Writing—Review & Editing, A.Z. and M.C.; Statistical Analysis, M.E.C.; Methodology, M.M.G.-M., B.G. and M.C.; Supervision and Project Administration, M.C. All authors have read and agreed to the published version of the manuscript.

Funding: This research was funded by the Spanish Ministry of Science and Innovation MCIN/AEI/10.13039/501100011033 through the grant RTI2018-098042-B100; and through the project SBPLY-19-180501-00074, which was co-financed by Junta de Comunidades de Castilla–La Mancha and the European Regional Development Fund. Guayente Latorre acknowledges a pre-doctoral contract (PREDUCLM-16215) from Universidad de Castilla–La Mancha and the European Social Fund. M. Mercedes García-Martínez acknowledges the grant for the requalification of the Spanish university system modality Margarita Salas (2022-NACIONA-11053), which was funded by the European Union–NextGeneration EU through the Universidad de Castilla–La Mancha. Miguel Ángel De la Cruz-Morcillo acknowledges a post-doctoral contract (2022-UNIVERS-11006) from Universidad de Castilla–La Mancha that was co-financed by the European Social Fund Plus (ESF+).

Data Availability Statement: The data presented in this study are available upon request from the authors.

Acknowledgments: The authors wish to thank Agroservicios Guayule S.L. and Nokian Tyres Plc. for their support in this research, as well as Martínez-Vázquez for kindly supplying of argentatins standards and Kenneth McCreath for proofreading the manuscript.

Conflicts of Interest: The authors declare no conflict of interest.

References

1. Rasutis, D.; Soratana, K.; McMahan, C.; Landis, A.E. A sustainability review of domestic rubber from the guayule plant. *Ind. Crops Prod.* **2015**, *70*, 383–394. [CrossRef]
2. Dehghanizadeh, M.; Mendoza Moreno, P.; Sproul, E.; Bayat, H.; Quinn, J.C.; Brewer, C.E. Guayule (*Parthenium argentatum*) resin: A review of chemistry, extraction techniques, and applications. *Ind. Crops Prod.* **2021**, *165*, 113410. [CrossRef]
3. Sfeir, N.; Chapuset, T.; Palu, S.; Lançon, F.; Amor, A.; García García, J.; Snoeck, D. Technical and economic feasibility of a guayule commodity chain in Mediterranean Europe. *Ind. Crops Prod.* **2014**, *59*, 55–62. [CrossRef]
4. Jara, F.M.; Cornish, K.; Carmona, M. Potential applications of guayulins to improve feasibility of guayule cultivation. *Agronomy* **2019**, *9*, 804. [CrossRef]
5. Teetor, V.H.; Ray, D.T.; Schloman, W.W. Evaluating chemical indices of guayule rubber content: Guayulins A and B. *Ind. Crops Prod.* **2009**, *29*, 590–598. [CrossRef]
6. Battistel, E.; Ramello, S.; Querci, C. Integrated process for processing and utilising the guayule plant. U.S. Patent 9969818 B2, 15 May 2018.
7. Jara, F.M.; Carrión, M.E.; Angulo, J.L.; Latorre, G.; López-Córcoles, H.; Zalacain, A.; Hurtado de Mendoza, J.; García-Martínez, M.M.; Carmona, M. Chemical characterization, antioxidant activity and morphological traits in the leaves of guayule (*Parthenium argentatum* A. Gray) and its hybrids. *Ind. Crops Prod.* **2022**, *182*, 114927. [CrossRef]
8. Cruz, V.M.V.; Dierig, D.A.; Lynch, A.; Hunnicutt, K.; Sullivan, T.R.; Wang, G.; Zhu, J. Assessment of phenotypic diversity in the USDA, National Plant Germplasm System (NPGS) guayule germplasm collection. *Ind. Crops Prod.* **2022**, *175*, 114303. [CrossRef]
9. Nabi, M.; Tabassum, N.; Ganai, B.A. Phytochemical screening and antibacterial activity of Skimmia anquetilia N.P. Taylor and Airy Shaw: A first study from Kashmir Himalaya. *Front. Plant Sci.* **2022**, *13*, 937946. [CrossRef]
10. Nabi, M.; Zargar, M.I.; Tabassum, N.; Ganai, B.A.; Wani, S.U.D.; Alshehri, S.; Alam, P.; Shakeel, F. Phytochemical Profiling and Antibacterial Activity of Methanol Leaf Extract of Skimmia anquetilia. *Plants* **2022**, *11*, 1667. [CrossRef]
11. Marcelino, S.; Mandim, F.; Taofiq, O.; Pires, T.C.S.P.; Finimundy, T.C.; Prieto, M.A.; Barros, L. Valorization of *Punica granatum* L. Leaves Extracts as a Source of Bioactive Molecules. *Pharmaceuticals* **2023**, *16*, 342. [CrossRef]
12. Seididamyeh, M.; Phan, A.D.T.; Sivakumar, D.; Netzel, M.E.; Mereddy, R.; Sultanbawa, Y. Valorisation of Three Underutilised Native Australian Plants: Phenolic and Organic Acid Profiles and In Vitro Antimicrobial Activity. *Foods* **2023**, *12*, 623. [CrossRef] [PubMed]
13. Haagen-Smit, A.J.; Siu, R. Chemical investigations in guayule. I. Essential oil of guayule, *Parthenium argentatum*, Gray. *J. Am. Chem. Soc.* **1944**, *66*, 2068–2074. [CrossRef]
14. Scora, R.W.; Kumamoto, J. Essential leaf oils of *Parthenium argentatum* A. Gray. *J. Agric. Food Chem.* **1979**, *27*, 642–643. [CrossRef]

15. Piana, F.; Ciulu, M.; Quirantes-Piné, R.; Sanna, G.; Segura-Carretero, A.; Spano, N.; Mariani, A. Simple and rapid procedures for the extraction of bioactive compounds from guayule leaves. *Ind. Crops Prod.* **2018**, *116*, 162–169. [CrossRef]
16. Piluzza, G.; Campesi, G.; Molinu, M.G.; Re, G.A.; Sulas, L. Bioactive compounds from leaves and twigs of guayule grown in a Mediterranean environment. *Plants* **2020**, *9*, 442. [CrossRef]
17. Rozalén, J.; García-Martínez, M.M.; Carrión, M.E.; Zalacain, A.; López-Córcoles, H.; Carmona, M. Effect of seasonal decrease in temperature on the content and composition of guayulins in stems of guayule (*Parthenium argentatum*, Gray). *Plants* **2021**, *10*, 537. [CrossRef]
18. Curtis, O.F. Distribution of rubber and resins in guayule. *Plant Physiol.* **1947**, *22*, 333–359. [CrossRef]
19. Gallego, B.; Carrión, M.E.; Latorre, G.; García-Martínez, M.M.; Hurtado de Mendoza, J.; Zalacain, A.; Carmona, M. UHPLC-HRMS profile and accumulation of argentatins in stems of 27 accessions of guayule (*Parthenium argentatum* A. Gray) and its hybrids along the growth cycle. *Ind. Crops Prod.* **2022**, *187*, 115463. [CrossRef]
20. Gallego, B.; García-Martínez, M.M.; Sánchez-Gómez, R.; Latorre, G.; de Mendoza, J.H.; Zalacain, A.; Carmona, M. Agronomic practices impact argentatin content in guayule (*Parthenium argentatum* A. Gray). *Ind. Crops Prod.* **2023**, *194*, 116402. [CrossRef]
21. Martínez-Vázquez, M.; Martínez, R.; Espinosa Pérez, G.; Díaz, M.; Herrera Sánchez, M. Antimicrobial properties of argentatin A, isolated from Parthenium argentatum. *Fitoterapia* **1994**, *65*, 371–372.
22. Flores-Rosete, G.; Martínez-Vázquez, M. Anti-inflammatory and cytotoxic cycloartanes from guayule (*Parthenium argentatum*). *Nat. Prod. Commun.* **2008**, *3*, 413–422. [CrossRef]
23. Alcántara-Flores, E.; Brechú-Franco, A.E.; García-López, P.; Rocha-Zavaleta, L.; López-Marure, R.; Martínez-Vázquez, M. Argentatin B inhibits proliferation of prostate and colon cancer cells by inducing cell senescence. *Molecules* **2015**, *20*, 21125–21137. [CrossRef] [PubMed]
24. Parra-Delgado, H.; Ramírez-Apan, T.; Martínez-Vázquez, M. Synthesis of argentatin A derivatives as growth inhibitors of human cancer cell lines in vitro. *Bioorganic Med. Chem. Lett.* **2005**, *15*, 1005–1008. [CrossRef] [PubMed]
25. Tavarez-Santamaría, Z.; Jacobo-Herrera, N.J.; Rocha-Zavaleta, L.; Zentella-Dehesa, A.; del Carmen Couder-García, B.; Martínez-Vázquez, M. A higher frequency administration of the nontoxic cycloartane-type triterpene argentatin A improved its anti-tumor activity. *Molecules* **2020**, *25*, 1780. [CrossRef] [PubMed]
26. Madasu, C.; Xu, Y.-M.; Wijeratne, E.M.K.; Liu, M.X.; Molnár, I.; Gunatilaka, A.A.L. Semi-synthesis and cytotoxicity evaluation of pyrimidine, thiazole, and indole analogues of argentatins A–C from guayule (*Parthenium argentatum*) resin. *Med. Chem. Res.* **2022**, *31*, 1088–1098. [CrossRef]
27. Romero-Benavides, J.C.; Bailon-Moscoso, N.; Parra-Delgado, H.; Ramirez, M.I.; Villacis, J.; Cabrera, H.; Gonzalez-Arevalo, G.; Cueva, R.; Zentella-Dehesa, A.; Ratovitski, E.A.; et al. Argentatin B derivatives induce cell cycle arrest and DNA damage in human colon cancer cells through p73/p53 regulation. *Med. Chem. Res.* **2018**, *27*, 834–843. [CrossRef]
28. Xu, Y.; Madasu, C.; Liu, M.X.; Wijeratne, E.M.K.; Dierig, D.; White, B.; Molnár, I.; Gunatilaka, A.A.L. Cycloartane- and lanostane-type triterpenoids from the resin of *Parthenium argentatum* AZ-2, a byproduct of guayule rubber production. *ACS Omega* **2021**, *6*, 15486–15498. [CrossRef] [PubMed]
29. Spano, N.; Meloni, P.; Idda, I.; Mariani, A.; Itria Pilo, M.; Marina, V.; Izabela Lachowicz, J.; Rivera, E.; Orona-Espino, A. Assessment, validation and application to real samples of a RP-HPLC method for the determination of guayulins A, B, C and D in guayule shrub. *Separations* **2018**, *5*, 23. [CrossRef]
30. Newman, D.J.; Cragg, G.M. Natural Products as Sources of New Drugs over the Nearly Four Decades from 01/1981 to 09/2019. *J. Nat. Prod.* **2020**, *83*, 770–803. [CrossRef]
31. Zhao, W.; Cong, Y.; Li, H.M.; Li, S.; Shen, Y.; Qi, Q.; Zhang, Y.; Li, Y.Z.; Tang, Y.J. Challenges and potential for improving the druggability of podophyllotoxin-derived drugs in cancer chemotherapy. *Nat. Prod. Rep.* **2021**, *38*, 470–488. [CrossRef]
32. Jackson, D.E.; Dewick, P.M. Tumour-inhibitory aryltetralin lignans from *Podophyllum pleianthum*. *Phytochemistry* **1985**, *24*, 2407–2409. [CrossRef]
33. Gallego, A.; Malik, S.; Yousefzadi, M.; Makhzoum, A.; Tremouillaux-Guiller, J.; Bonfill, M. Taxol from *Corylus avellana*: Paving the way for a new source of this anti-cancer drug. *Plant Cell. Tissue Organ Cult.* **2017**, *129*, 1–16. [CrossRef]
34. Rozalén, J.; García-Martínez, M.M.; Carrión, M.E.; Carmona, M.; López-Córcoles, H.; Cornish, K.; Zalacain, A. Adapting the accelerated solvent extraction method for resin and rubber determination in guayule using the BÜCHI speed extractor. *Molecules* **2021**, *26*, 183. [CrossRef]

plants

MDPI

Review

Functional, Chemical, and Phytotoxic Characteristics of *Cestrum parqui* L'Herit: An Overview

Maria Chiara Di Meo [1], Cinzia Di Marino [2], Pasquale Napoletano [3,*], Anna De Marco [4], Anna Rita Bianchi [5], Silvana Pedatella [2] and Domenico Palatucci [5,*]

1 Department of Sciences and Technologies, University of Sannio, 82100 Benevento, Italy; mardimeo@unisannio.it
2 Department of Chemical Sciences, University of Naples Federico II, 80126 Naples, Italy; cdimarin@unina.it (C.D.M.); silvana.pedatella@unina.it (S.P.)
3 Department of Agriculture, Food, Environment and Forestry (DAGRI), University of Florence, 50144 Florence, Italy
4 Department of Pharmacy, University of Naples Federico II, 80131 Naples, Italy; ademarco@unina.it
5 Department of Biology, University of Naples Federico II, 80126 Naples, Italy; annarita.bianchi@unina.it
* Correspondence: pasquale.napoletano@unifi.it (P.N.); domenico.palatucci@unina.it (D.P.)

Abstract: *Cestrum parqui* L'Herit. (Solanaceae family) is a species of forest shrub, self-incompatible and specialized in pollination, widespread in the subtropical area of the planet, and now widely distributed also in the Mediterranean area. The constituents of its leaves have antimicrobial, anticancer, insecticidal, antifeedant, molluscicidal, and herbicidal properties. The spread of this species represents a valuable source of compounds with high biological value. Various research groups are engaged in defining the chemical composition of the different parts of the plant and in defining its properties in view of important and promising commercial applications. To date, there are only a few incomplete reports on the potential applications of *C. parqui* extracts as selective natural pesticides and on their potential phytotoxic role. Scientific knowledge and the use of extraction techniques for these components are essential for commercial applications. This article summarizes the research and recent studies available on the botany, phytochemistry, functional properties, and commercial applications of *C. parqui*.

Keywords: *Cestrum parqui* L'Herit.; insecticidal and antifeedant activity; herbicidal activity; secondary metabolites; lignans; flavones; oxylipins

1. Introduction

Plants synthesize numerous secondary metabolites through specific metabolic pathways [1]. These molecules are interesting for plant ecology, reproduction, and physiology. The classes and types of secondary metabolites produced are a useful tool in phylogenetic and taxonomic studies [2]. Humans have exploited natural substances for a long time in medicinals, agriculture, arts, food, feed, and religion. Specifically, the application of natural products in agriculture is nowadays leveraged in order to reduce human's cultures impact on ecosystems and on public health [3]. This review is focused on *Cestrum parqui* L'Herit's (green cestrum, Figure 1) secondary metabolites content and the effects of plant extracts and purified chemicals in pest and weeds controlling. This plant belongs to the Solanaceae family and has been cultivated throughout the world in gardens as an ornamental species. *Cestrum parqui* is also known as Chilean cestrum, Chilean flowering jessamine, Chilean jessamine, green cestrum, green poison berry, green poison-berry, green poisonberry, iodine bush, willow jasmine, willow leaved jessamine, willow-leaf jessamine, willow-leaved jasmine, and willow-leaved jessamine [4]. It was been first described by Carolus Ludovicus L'Heritier in 1785 [5], and its origins trace back to South America [6]. Nowadays, it is widely diffused all over the world, and classified according to the new systematic [7,8].

Figure 1. Flowers of *Cestrum parqui* L'Herit.

Taxonomy, Morphology, and Distribution

C. parqui belongs to the Solanaceae family, a family of dicotyledonous angiosperms largely cultivated by humans as horticultural—*Solanum tuberosum*, *Solanum melongena*, *Solanum lycopersicum*, and *Capsicum annuum*, medicinal—*Atropa belladonna*, and recreational crops—*Nicotiana tabacum*. The family comprises 100 genera with about 2500 species. The genus that hosts the largest number of plants is *Solanum*, with about 1400 species; conversely, *Licianthes* includes 200 species, and *Cestrum* has 175 species [9,10]. Solanaceae are represented in the wild on all continents, with a greater number of species in the American continent, and they adapt well to almost all ecosystems, although most of them prefer warmth rather than intense cold. *C. parqui* is native to Brazil, Bolivia, northern and central Chile, Peru, Paraguay, Uruguay, and northern and central Argentina, but, today, it is widely spread in southeastern and eastern parts of Australia, New Zealand, in some parts of the southern United States of America like California and Texas, and in much of Europe (Figure 2) [11].

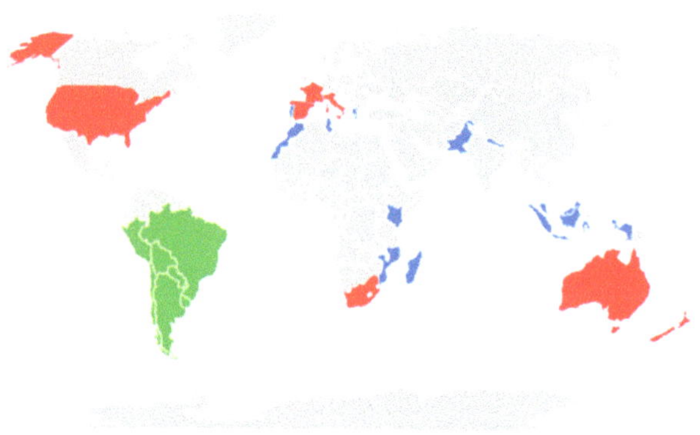

Figure 2. World distribution of *C. parqui*.

This plant is one of the main weeds of the Mediterranean area [12]. Its spread began as an ornamental plant, but it soon became invasive in many warm temperate and subtropical regions because it adapts well to the edges of watercourses and is also found in parks, old gardens, uncultivated areas, open woods, forest edges, pastures, and along roadsides. In Australia, *C. parqui* is considered an environmental weed, meaning it has no agricultural function, damaging and competing with existing plants, especially in New South Wales and Queensland [13]. For this reason, it is currently listed as a priority environmental weed in three regions and a sleeper weed in other parts of the country [14]. The invasiveness of this plant is particularly evident when it forms dense stands along forest edges and watercourses, replacing native plants in these habitats and preventing their regeneration [13]. It is an erect, highly branched shrub that usually grows 1–3 m tall, but occasionally reaches up to 5 m in height. It has tubular flowers in clusters, yellow or greenish-yellow in color, and stems and leaves that have an unpleasant odor when crushed. The taxonomy of the plant is described in Table 1.

Table 1. Phylogenetic taxonomy of *C. parqui* [4,9,10].

Kingdom	Plantae
Sub-kingdom	Tracheophytes
Division	Angiosperms
Class	Eudicots
Sub-class	Asterids
Order	Lamianae Takht
Sub-order	Solanales Juss. ex Bercht. and J. Presl
Family	Solanaceae' Juss
Tribe	Cestreae
Genus	*Cestrum* L.
Species	*parqui*

2. Traditional Use and Properties

In Chilean folk medicine, it was used as an antipyretic and for the treatment of fever and inflammation [15,16]. Extracts of the plant obtained with solvents of different polarity have shown moderate antimicrobial activity against the fungi *Penicillium expansum* and *Candida albicans* and the bacteria *Bacillus subtilis*, *Pseudomonas aeruginosa*, *Staphylococcus aureus*, *Escherichia coli*, and *Streptococcus pneumoniae* [17]. The methanolic extract of the leaves has exhibited possible anticancer activity on the human myeloid leukemia cell line (HL-60) and an antiproliferative effect on another two cell lines (HT-29 and Molt-3 cells); this may be due to the presence of ursolic and oleanolic acids, two pentacyclic triterpenes [18,19], as well as the ability to inhibit platelet aggregation induced by ADP and/or collagen, both in sheep and human blood [20]. Furthermore, the methanolic extract of *C. parqui* leaves has a strong effect on sperm motility in vitro. Electron microscopy studies on human sperm, incubated with concentrations ranging from 40 to 250 μg/mL of *C. parqui* leaf extract and at time intervals ranging from 5 to 240 min, have shown damage to the head and acrosomal membranes, with a maximum spermicidal effect at the highest tested concentration, generally dose- and time-dependent [21]. A screening among *Cestrum* spp. reveals *Aspergillus terreus* as an endophytic fungus of *C. parqui* leaves. This microbe biosynthesizes camptothecin, a modified monoterpene indole alkaloid used in cancer chemotherapy [22,23].

3. Potential Effects of *C. parqui*

3.1. Insecticidal and Antifeedant Activity

The genus *Cestrum* is rich in saponins, and most species exhibit toxicity that supports their use as potential insecticides, herbicides, molluscicides, antimicrobial agents, and antitumor agents. In fact, as far back as the early 1950s, the discovery of gitogenin and digitogenin in the green berries of *C. parqui* [24], or tigogenin and digallogenin in dried leaves, has been documented [25].

The leaves of *C. parqui* are the most studied organ of the plant. This is likely related to the observation that many animals that had eaten its leaves were severely intoxicated. For example, several cases of cattle poisoning occurred in Chile between 1992 and 1998 and in Brazil starting from the late 1960s [26]. Necropsies of the animals showed pulmonary edema, congestion, hemorrhages in various organs, and hepatic dysfunction [27]. A few years later, Babouche et al. [28] demonstrated that the saponin-rich fraction obtained from the hydroalcoholic extract of the plant could interfere with insect metabolism by lowering the amount of cholesterol needed for ecdysone production, a molting hormone [29].

C. parqui's insecticidal activity has been tested on different species. Specifically, aqueous extracts of *C. parqui* have been evaluated on *Ceratitis capitata*, commonly known as the Mediterranean fruit fly, at different concentrations [30]. *C. capitata*, widespread in

Africa, the Mediterranean basin, and South America, is a highly polyphagous species whose larvae develop in a wide range of fruits and are responsible for significant economic damage to the agricultural sector [31]. Water and organic solvent plant extracts have been tested. The aqueous extract at 0.6% (*w*/*w*) of the plant completely inhibited the pupation process of the neonate larvae, while extracts obtained with organic solvents were almost harmless. In another experiment, ethanolic and water extracts obtained from young and old leaves were tested on *Xanthogaleruca luteola* adult insects. In this case, higher mortality has been obtained for ethanolic extract of young leaves [32]. The leaves are likewise active against desert locust *Schistocerca gregaria*. Some of the contained chemicals interfere with the exuviation process, causing insect death at different evolutionary stages [33].

These results were confirmed a few years later by Chaieb et al. [34,35]. The authors evaluated the entomotoxic activity on *Schistocerca gregaria*, a polyphagous and voracious grasshopper that feeds on leaves, flowers, shoots, fruits, and seeds of various plant species, including numerous species of primary importance to humans such as rice, barley, corn, sorghum, sugarcane, cotton, date palm, and banana; *Spodoptera littoralis*, an insect that can attack numerous economically important crops such as turnips, tomatoes, hemp, hibiscus, purslane, mint, clover, tobacco, mallow, apple, grapevine, and many others; *Tribolium confusum*, an insect that mainly feeds on natural products such as cereals and flour, rice, dried fruit, powdered milk, mouse baits, spices, and corn; and *Culex pipiens*, the most common mosquito in the Northern Hemisphere, hematophagous and harmful to health. Chaieb and colleagues [34,35] performed toxicity tests based on the species, through simple contact, injection, forced ingestion, or addition to the food substrate. In the case of contact tests, the results were modest, probably because the saponins were unable to penetrate the waxy cuticle of the target organisms, evidently due to their hydrophilicity; while in ingestion tests, the food substrate was probably unpalatable to the target animal. The best results were obtained with injection, which is obviously impractical in daily practice. However, the results show greater activity on *S. littoralis* followed by *C. pipiens*, and slightly less on *S. gregaria* and *T. confusum*. In any case, the chances of using the crude material as it is, to be added to the diet of the target organism, seem slim. Apparently, the added product had lost its palatability, suggesting the need to isolate the saponins present in the crude material for individual use. It is not disregarded that the problem can be overcome by delivering the saponins through softer and more palatable foods preferred by the target insects.

The antifeedant effect of the aqueous extract of *C. parqui* has also been measured on *Pieris brassicae*, a butterfly that mainly feeds on cultivated varieties of Brassicaceae, especially *Brassica oleracea* (cabbage), and plants of the genus *Tropaeolum* [36]. The effect of increasing amounts of extract, added in percentages of 2, 4, 8, 16, and 32% to the lepidopteran's diet, was measured, showing a delay in larval growth at lower concentrations, abnormal metamorphosis at intermediate concentrations, and death at the highest concentration.

It is interesting to note that it has been proven that the activity significantly decreases with the loss of the sugar bound to the steroid nucleus [37], much like what happens in the case of α-chaconine and α-solanine. Thus, it is not surprising that the saponins of *C. parqui* are completely ineffective against the phytopathogenic fungi *Fusarium solani* and *Botrytis cinerea*, which probably can secrete detoxifying enzymes capable of hydrolyzing the sugar chains [38].

For economic and especially environmental reasons (reassuring an increasingly reluctant civil society to the use of chemical products), many research groups are committed to identifying specific natural insecticides. For example, pine wood is particularly susceptible to colonization by organisms of the genus *Leptographium* spp. *Ophiostoma* and *Ceratocystis*, which, by invading the vessels, block the passage of sap and cause deterioration phenomena, including the death of plants. The activity of these pathogens, which is only visible by stripping the trunks of dead or suffering plants, is preceded by that of their vectors, mostly coleopteran insects like *Hylurgus ligniperda*, which, by invading plants weakened by various stresses, let the fungus penetrate the sapwood, following their galleries. The symptoms of the infection consist of a wilting of the canopy, with needles quickly turning from pale green to brown and drying up. Wood assortments undergo considerable depreciation due to aesthetic defect. To date, *H. ligniperda* is controlled using methyl iodide, a chemical product dangerous for users and the environment. Huanquilef et al. [39] tested various fractions of the ethanolic extract of *C. parqui* leaves on *H. ligniperda*. Thus, the fractions soluble in chloroform, ethyl acetate, and butanol, which were then added to the insect's diet, were considered. Over the course of seven days, it was demonstrated that the considered extracts influenced the feeding behavior of the target organism in both adult organisms and larvae, with a dose-dependent effect. In particular, the chloroform extract was the most active, even at low concentrations (0.4% w/v), indicating that it can be considered for an economical and relatively simple commercial application.

3.2. Molluscicidal Activity

In the past twenty years, several authors who have tested the insecticidal and antifeeding activity of hydroalcoholic extracts from leaves have also tested their molluscicidal activity, using the snail *Theba pisana* as a target organism [34,35,40], a gastropod mollusk introduced into numerous areas including northern Europe, North America, parts of Africa, Asia, and Australia, where it has often become an invasive species, posing a serious problem for agriculture. These extracts are mostly produced from leaves dried at 40 °C for four days and then finely powdered. They are first extracted with petroleum ether to remove fats and then with methanol. The methanolic extract is then washed with diethyl ether, causing the precipitation of the saponin-containing fraction, which is used for various types of experiments. In some experiments, the reaction mixture was deposited at the bottom of containers where the snails were free to move, or the mixture was applied directly to the bodies of the target organisms. These snails responded to the presence of the saponin-containing mixture with a strong production of mucus, which caused their dehydration and, ultimately, death. In other experiments, the saponin-containing mixture was added to corn bran or deposited on cabbage leaves, which the snails normally feed on, or was dissolved in water in varying amounts from 2 to 8 mg/mL. In these latter three cases, the effects were minimal and mostly reduced to a slight weight loss in the animals, which stopped eating or drinking, evidently recognizing the presence of the toxic substance [40]. However, it is likely that administering the saponin mixture in different and more palatable foods could yield better results than simple contact. Considering the good results obtained, experimentation on *T. pisana* continued with the aim of understanding if there were differences in the toxicity of the saponin-rich crude material on the juvenile or adult form of the mollusk [41]. Two tests lasting 24 h each were used, and each was repeated three times. In the first test, the saponin-containing fraction was deposited on the surface where the snails moved at concentrations of 10, 100, 500, 1000, and 2000 ppm, respectively. A mortality rate of 100% was found with a concentration of the analyzed fraction equal to 315 $\mu g/cm^2$ for adults and 157 $\mu g/cm^2$ for juveniles, with LD_{50} values of 36 and 6 ppm, respectively. In the second test, the saponin-containing fraction was placed in direct contact with the back of the target organism in quantities of 1, 5, and 10 mg, respectively. A mortality rate of 100% was found with a crude quantity of 10 mg for adults and 5 mg for juveniles, with LD_{50} values of 2.6 and 1.0 mg/animal, respectively.

3.3. Phytochemical Composition

The bark and, especially, the leaves of this shrub contain a large number of secondary metabolites, generally of low molecular weight. These metabolites have been isolated from polar infusions obtained via extraction with methanol and/or ethanol/water (Scheme 1), purified using direct and reverse-phase chromatographic techniques (on column or HPLC), and identified using spectroscopic techniques (NMR, both one-dimensional and two-dimensional) and mass spectrometry. Specifically, from the organic infusions, C_{13}-nor-isoprenoids (**1–18**, Figure 3), sesquiterpenes (**19–20**, Figure 4), a spirostane (**21**, Figure 4), a pseudosapogenin (**22**, Figure 4), lignans (**23–42**, Figure 5), aromatic compounds (**43–47** and **89**, Figure 6), oxylipins (**63–76**, Figure 8), kaurenic glycosides (**77–78**, Figure 9), steroid saponins (**79–86**, Figure 10), an aromatic glycoside (**89**, Figure 10), and pentacyclic triterpenes (**90–91**, Figure 10) have been isolated. Conversely, from the aqueous infusion, aromatic compounds (**48–58**, Figure 6), flavones (**60–62**, Figure 7), a steroid saponin (**87**, Figure 10), and a glycoalkaloid (**88**, Figure 10) have been isolated.

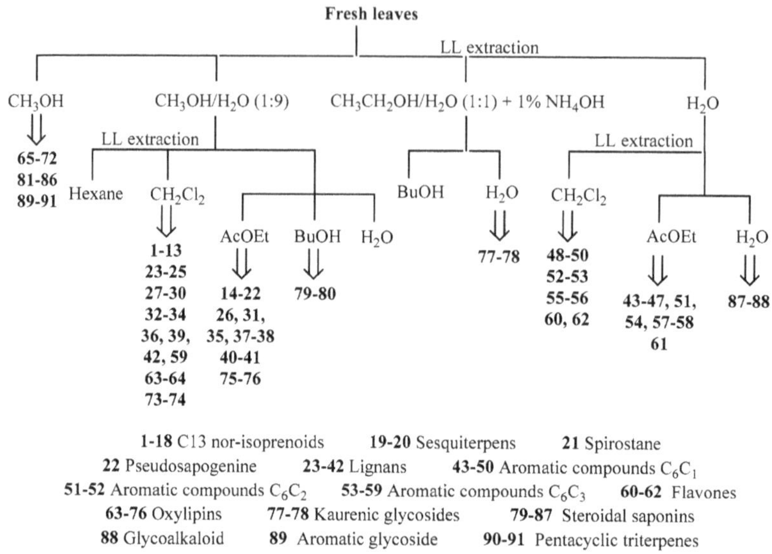

Scheme 1. Isolation of C_{13}-nor-isoprenoids, sesquiterpenes, spirostanes, pseudosapogenins, lignans, aromatic compounds, flavones, and oxylipins.

3.4. Herbicidal Activity

From the fresh leaves of *C. parqui*, a total of 76 compounds were isolated (Table 2), which after purification and structural determination were tested to evaluate their phytotoxic activity. In particular, the fresh leaves were finely chopped and then infused with methanol, methanol/water: 9/1 (*v*/*v*), water/ethanol: 1/1 (*v*/*v*) + 1% NH_4OH, and water, respectively. The last three infusions were dried and then extracted with solvents of increasing polarity as indicated in Scheme 1. Thus, after numerous chromatographic steps, the following compounds were isolated: the C_{13}-nor-isoprenoids [42,43]; the sesquiterpenes [43]; the spirostan [43]; the pseudosapogenin [43]; the lignans [44,45]; the aromatic compounds 43–59 [46,47]; the flavones [46]; and the oxylipins [48].

Table 2. Secondary metabolites isolated from the leaves of *C. parqui* and tested for their potential phytotoxic activity in different studies.

No.	Name	Ref
C_{13}-nor-isoprenoids		
1	(6R,9R)-9-Hydroxy-4-megastigmen-3-one	
2	(2R,6R,9R)-2,9-Dihydroxy-4-megastigmen-3-one	
3	Byzantionoside A	
4	Annuionone E	
5	(3R,6R,7E,9R)-3,9-Dihydroxy-4,7-megastigmadiene	
6	(3R,6R,7E)-3-Hydroxy-4,7-megastigmadien-9-one	
7	(6R,7E,9R)-9-Hydroxy-4,7-megastigmadien-3-one	
8	Byzantionoside B	
9	Corcoionolo C	
10	(3S,5R,6R,7E,9R)-3,5,6,9-Tetrahydroxy-7-megastigmene	[42,46]
11	(3S,5R,6R,7E,9R)-5,6,9-Trihydroxy-3-isopropyloxy-7-megastigmene	
12	4-Oxo-β-ionol/(7E)-9-Hydroxy-5,7-megastigmadien-4-one	
13	(3S,7E,9R)-3,9-Dihydroxy-5,7-megastigmadiene	

Table 2. *Cont.*

No.	Name	Ref
14	(3R,7E)-3-Hydroxy-5,7-megastigmadien-9-one	
15	(3S,5R,6S,7E)-5,6-Epoxy-3-hydroxy-7-megastigmen-9-one	
16	(3S,5R,6S,7E,9R)-5,6-Epoxy-3,9-dihydroxy-7-megastigmene	
17	(6E,9S)-9-Hydroxy-4,6-megastigmadien-3-one	
18	(6Z,9S)-9-Hydroxy-4,6-megastigmadien-3-one	
Sesquiterpenes		
19	12-Hydroxy-α-cyperone	[43]
20	1,2,2a,3,6,7,8,8a-Octahydro-7-hydroxy-2a,7,8-trimethylacenaphthylen-4(4H)-one	[49,50]
Spirostane		
21	5α-Spirostan-3β,12β,15α-triol	[43]
Pseudosapogenine		
22	26-O-(30-Isopentanoyl)-β-D-glucopyranosyl-5α-furost-20(22)-ene-3β,26-diol	[43]
Lignans Diepoxylignans		
23	(+)-Pinoresinol	[44]
24	(+)-Mediaresinol	
25	(+)-Syringaresinol	
26	β-D-Glucopyranoside, 2,6-dimethoxy-4-[(1S,3aR,4S,6aR)-tetrahydro-4-(4-hydroxy-3,5-dimethoxyphenyl)-1H,3H-furo[3,4-c]furan-1-yl]phenyl	
Epoxylignans		
27	(+)-Lariciresinol	[44]
28	(+)-Justiciresinol	
29	5′-Methoxylariciresinol	
30	(−)-Berchemol	
Neolignans		
31	cis-Dehydrodiconiferyl alcohol	[44]
32	*trans*-Dehydrodiconiferyl alcohol	
33	(−)-Simulanol,	
Sesquilignans		
34	rel-(7Z,7′β,7″β,8′α,8″α)-4″,9,9′,9″-Tetrahydroxy-3,3′,3″-trimethoxy-4,7′:4′,7″-diepoxy-5,8′:5′,8″-sesquilign-7-ene	[44]
35	Herpetotriol	
36	rel-(7Z,7′α,7″α,8′β,8″β)-4″,9,9′,9″-Tetrahydroxy-3,3′,3″-trimethoxy-4,7′:4′,7″-diepoxy-5,8′:5′,8″-sesquilign-7-ene	
37	rel-(7E,7′α,7″α,8′β,8″β)-4″,9,9′,9″-Tetrahydroxy-3,3′,3″-trimethoxy-4,7′:4′,7″-diepoxy-5,8′:5′,8″-sesquilign-7-ene	
40	threo-4′,4″,7″,9″-Tetrahydroxy-3,3′,3″,5′-tetramethoxy-4,8″-oxy-7,9′:7′,9-diepoxylignan	
41	erythro-4′,4″,7″,9″-Tetrahydroxy-3,3′,3″,5′-tetramethoxy-4,8″-oxy-7,9′:7′,9-diepoxylignan	
42	erythro-4′,4″,7″,9″-Tetrahydroxy-3,3′,3″,5,5′-pentamethoxy-4,8″-oxy-7,9′:7′,9-diepoxylignan	
Oxyneolignan		
38	Dimethyl (7′E)-3,3′-dimethoxy-4,40-oxyneolign-7′-ene-9,9′-dioate	[44,45]
Norlignan		
39	9′-Nor-3′,4,4′-trihydroxy-3,5-dimethoxylign-7-eno-9,7′-lactone	[44,45]
Aromatic compounds		
43	4-Hydroxybenzaldehyde	[46,47,51–53]
44	3,5-Dimethoxybenzaldehyde	
45	4-Hydroxybenzoic acid	
46	Vanillic acid	
47	Syringic acid	
48	Methyl 4-hydroxybenzoate	
49	Methyl vanillate	
50	Methyl syringate	
51	Tirosol	
52	3′,5′-Dimethoxy-4′-hydroxy-2-hydroxy-acetophenone	

Table 2. *Cont.*

No.	Name	Ref
53	p-Coumaric acid	
54	Caffeic acid	
55	Methyl ferulate	
56	Methyl ester of caffeic acid	
57	p-Dihydrocoumaric acid	
58	Methyl ester of p-dihydrocoumaric acid	
59	Dihydrosynapic acid	
Flavones		
60	4′-Hydroxy-4-methoxychalcon	
61	Quercitin	[46]
62	N-(p-Carboxymethylphenyl)-p-hydroxybenzamide	
Oxylipins		
63	(8S,9R,10E,12R,14Z)-Heptadeca-10,14-diene-1,8,9,12-tetraol	
64	(8S,9R,10E,12S,14Z)-Heptadeca-10,14-diene-1,8,9,12-tetraol	
65	(9S,10R,11E,13R,15Z)-9,10,13-Trihydroxyoctadeca-11,15-dienoic acid	
66	(9S,10R,11E,13S,15Z)-9,10,13-Trihydroxyoctadeca-11,15-dienoic acid	
67	Methyl (9S,10R,11E,13R,15Z)-9,10,13-trihydroxyoctadeca-11,15-dienoate	
68	Methyl (9S,10R,11E,13S,15Z)-9,10,13-trihydroxyoctadeca-11,15-dienoate	
69	(9S,10R,11E,13R)-9,10,13-Trihydroxyoctadec-11-enoic acid	
70	(9S,10R,11E,13S)-9,10,13-Trihydroxyoctadec-11-enoic acid	[48]
71	Methyl (9S,10R,11E,13R)-9,10,13-trihydroxyoctadec-11-enoate	
72	Methyl (9S,10R,11E,13S)-9,10,13-trihydroxyoctadec-11-enoate	
73	(8S,9S,10R,11Z,14Z)-Heptadeca-11,14-diene-1,8,9,10-tetraol	
74	(9S,10S,11R,12Z,15Z)-9,10,11-Trihydoxyoctadeca-12,15-dienoic acid	
75	Methyl 10-hydroxyoctadec-12-enoate	
76	Methyl (9S,10E,12Z,15Z)-octadeca-10,12,15-trien-9-ol	

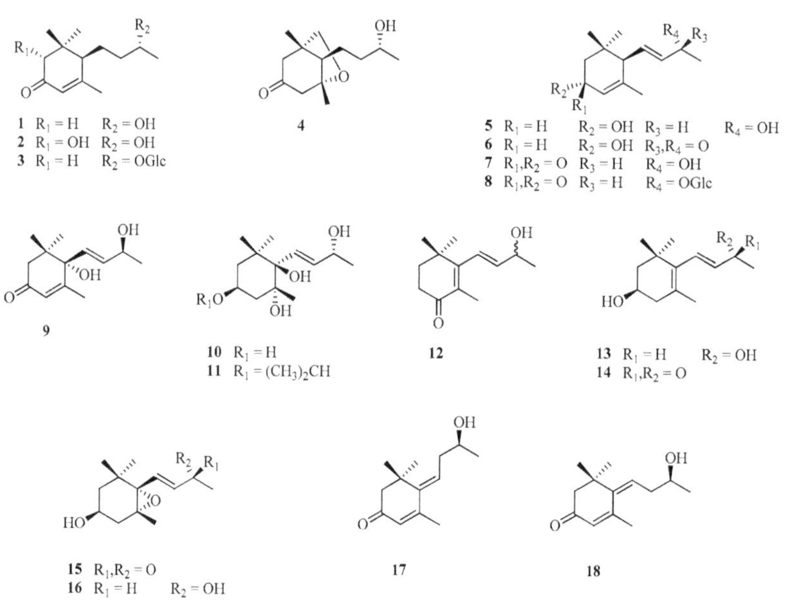

Figure 3. C_{13}-nor-isoprenoids isolated from *C. parqui*.

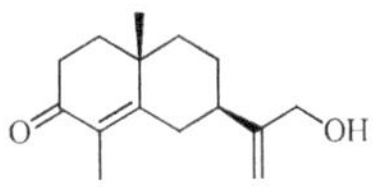

19

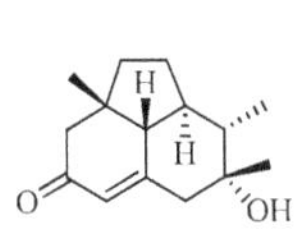

20

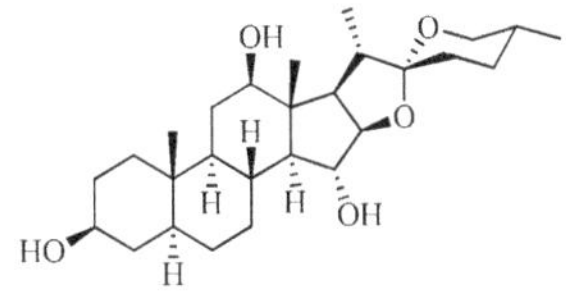

21

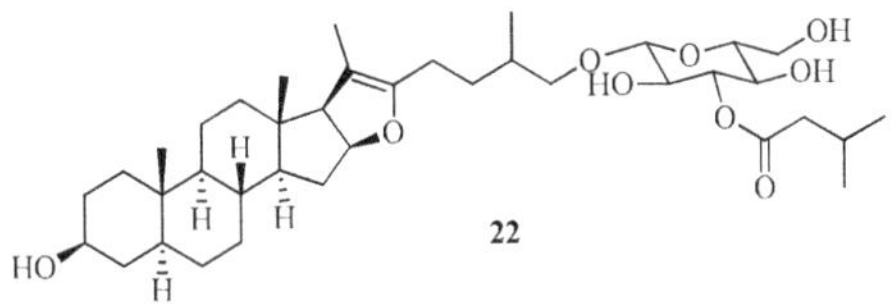

22

Figure 4. Sesquiterpenes (**19–20**), spirostane (**21**), and pseudosapogenin (**22**) isolated from *C. parqui*.

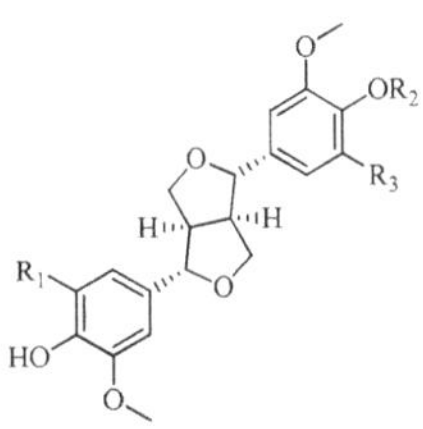

23 $R_1 = R_2 = R_3 = H$
24 $R_1 = R_2 = H$ $R_3 = OCH_3$
25 $R_1 = R_3 = OCH_3$ $R_2 = H$
26 $R_1 = R_3 = OCH_3$ $R_2 = Glc$

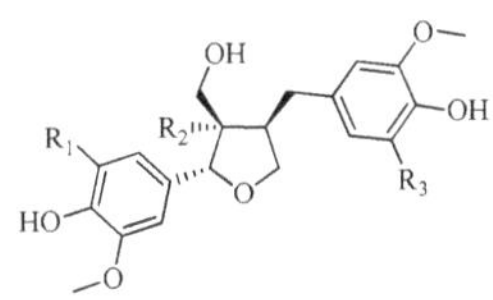

27 $R_1 = R_2 = R_3 = H$
28 $R_1 = R_2 = H$ $R_3 = OCH_3$
29 $R_1 = OCH_3$ $R_2 = R_3 = H$
30 $R_1 = R_3 = H$ $R_2 = OH$

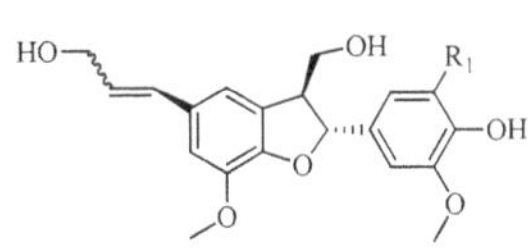

31 *cis* $R_1 = H$
32 *trans* $R_1 = H$
33 *trans* $R_1 = OCH_3$

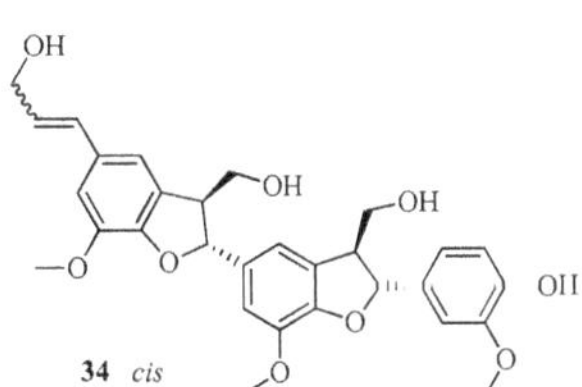

34 *cis*
35 *trans*

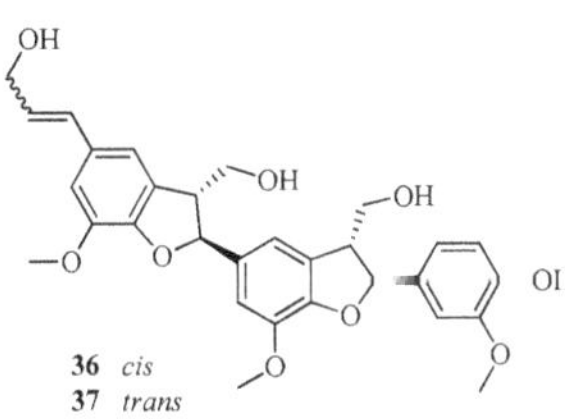

36 *cis*
37 *trans*

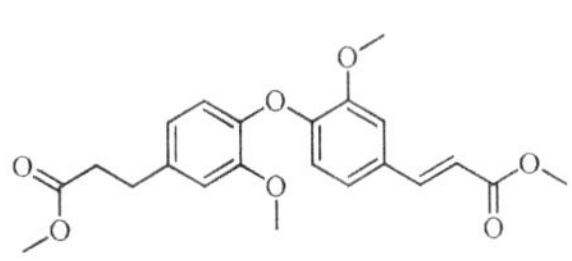

38

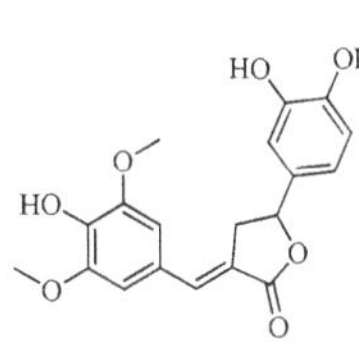

39

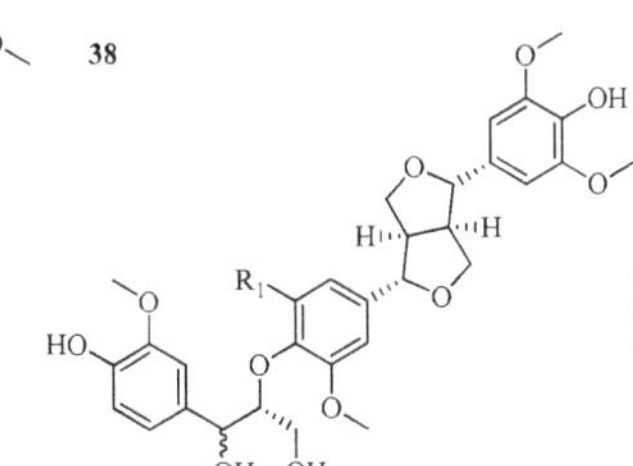

40 *threo* $R_1 = H$
41 *erythro* $R_1 = H$
42 *erythro* $R_1 = H$

Figure 5. Lignans from *C. parqui*.

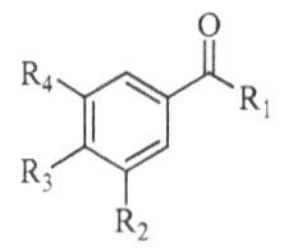

43 $R_1 = R_2 = R_4 = H$ $R_3 = OH$
44 $R_1 = R_3 = H$ $R_2 = R_4 = OMe$
45 $R_1 = R_3 = OH$ $R_2 = R_4 = H$
46 $R_1 = R_3 = OH$ $R_2 = OMe$; $R_4 = H$
47 $R_1 = R_3 = OH$ $R_2 = R_4 = OMe$
48 $R_1 = OMe$ $R_2 = R_4 = H$; $R_3 = OH$
49 $R_1 = R_2 = OMe$ $R_3 = OH$; $R_4 = H$
50 $R_1 = R_2 = R_4 = OMe$ $R_3 = OH$

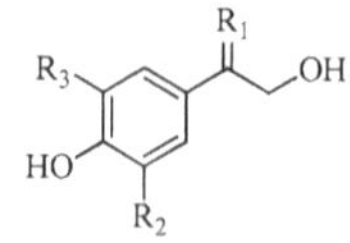

51 $R_1 = H,H$; $R_2 = R_3 = H$
52 $R_1 = O$; $R_2 = R_3 = OMe$

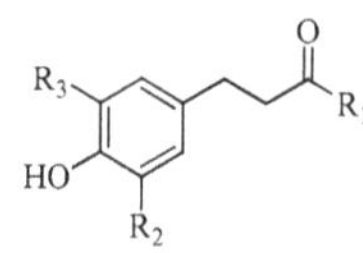

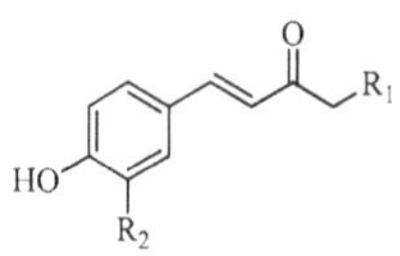

53 $R_1 = OH$; $R_2 = H$
54 $R_1 = R_2 = OH$
55 $R_1 = R_2 = OMe$
56 $R_1 = OMe$; $R_2 = OH$

57 $R_1 = OH$; $R_2 = R_3 = H$
58 $R_1 = OH$; $R_2 = OMe$; $R_3 = H$
59 $R_1 = R_2 = R_3 = OMe$

Figure 6. Aromatic compounds C_6C_1 (**43–50**), C_6C_2 (**51–52**), and C_6C_3 (**53–59**) from *C. parqui*.

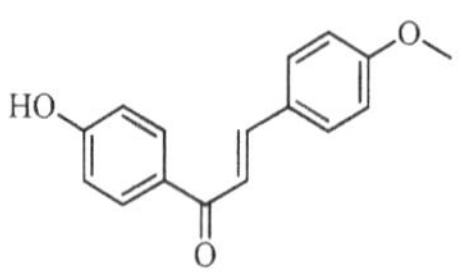

60

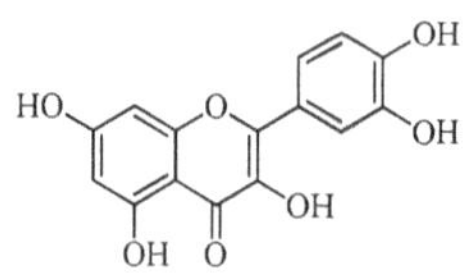

61

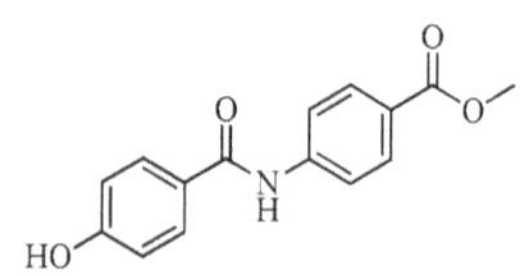

62

Figure 7. Flavonoids from *C. parqui*.

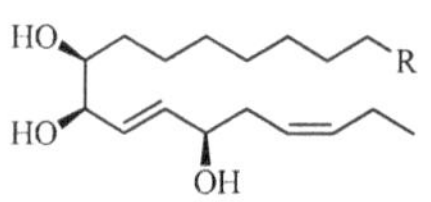

63 R = OH
65 R = COOH
67 R = COOMe

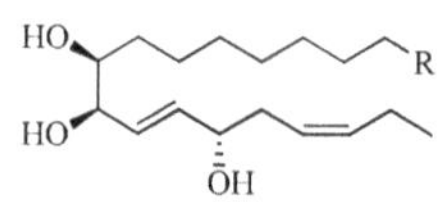

64 R = OH
66 R = COOH
68 R = COOMe

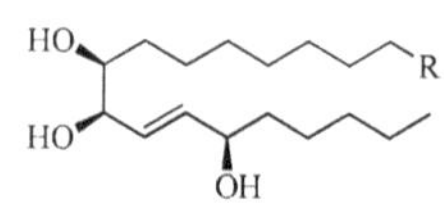

69 R = COOH
71 R = COOMe

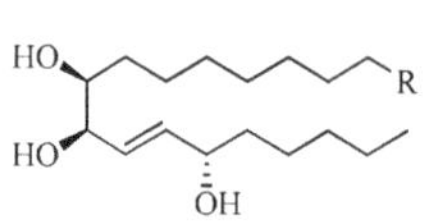

70 R = COOH
72 R = COOMe

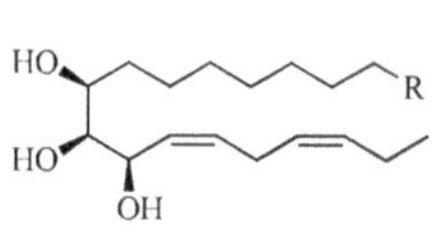

73 R = OH
74 R = COOH

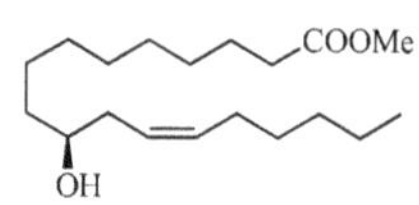

75

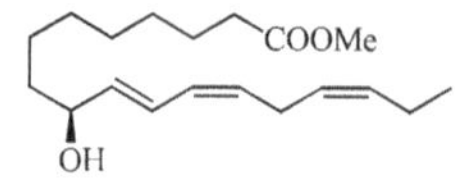

76

Figure 8. Polyol fatty acids **63–64** and **73**, methylated oxylipins **67** and **68**, and oxylipins **65–66**, **69–72**, and **74–76** (Note that **63/64**, **65/66**, **67/68**, **69/70**, and **71/72** are five diastereomeric couples of compounds).

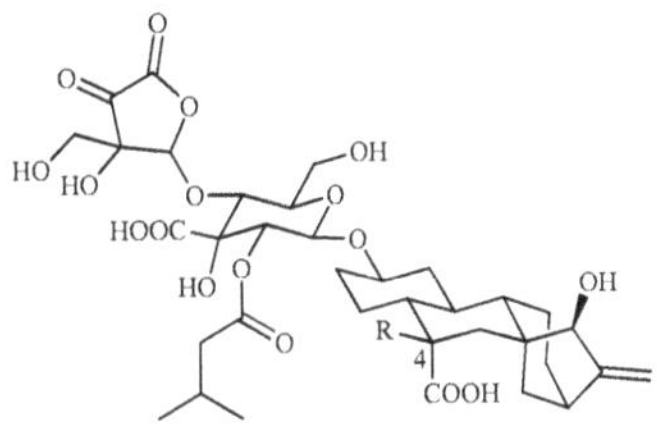

77 R = COOH Carboxyparquin
78 R = H Parquin

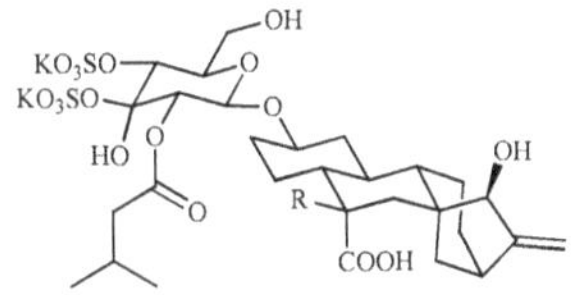

A R = COOH Carboxyactractyloside
B R = H Actractyloside

Figure 9. Kaurenic glycosides (**77–78**).

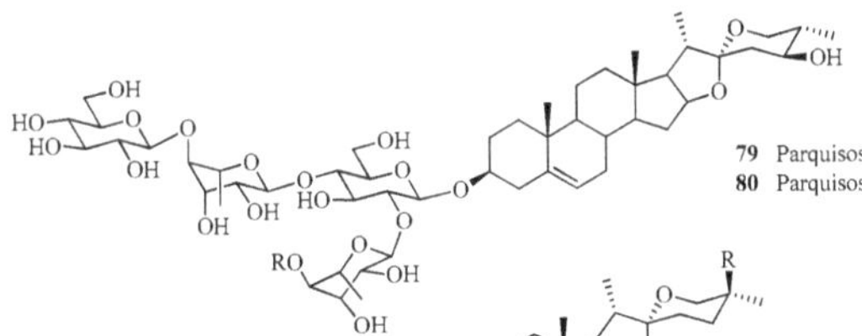

79 Parquisoside A R = H
80 Parquisoside B R = α-*L*-Rhamnopyranosyl

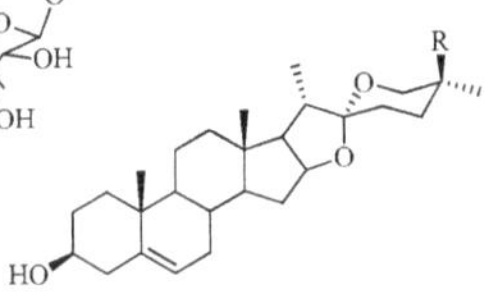

81 Neotigogenin R = OH
82 (25*R*)-Isonuatigenin R = OH

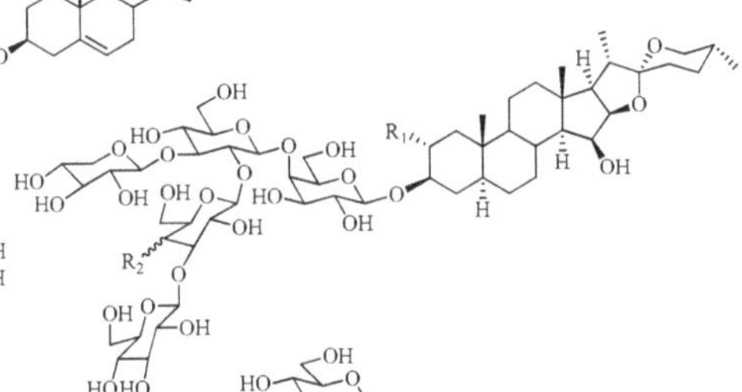

83 Parquispiroside R_1 = H R_2 = αOH
87 Digitogenin R_1 = OH R_2 = βOH

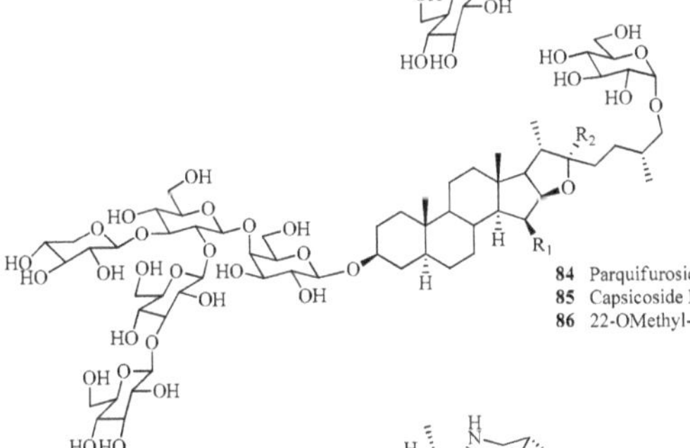

84 Parquifuroside $R_1 = R_2$ = OH
85 Capsicoside D R_1 = H R_2 = OH
86 22-OMethyl-capsicoside D R_1 = H R_2 = OMe

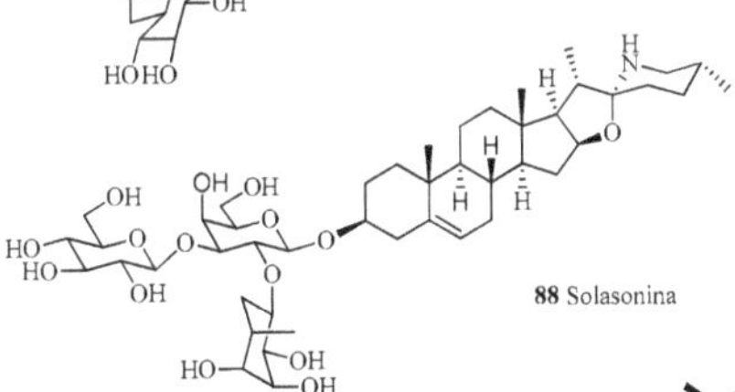

88 Solasonina

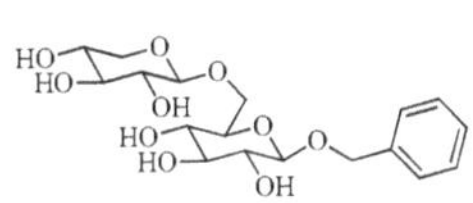

89 Benzil primeveroside

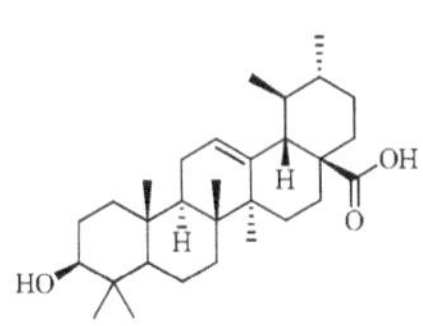

90 Ursolic acid

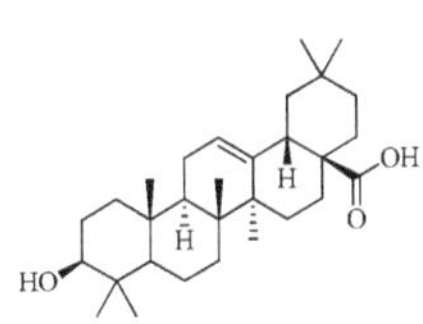

91 Oleanolic acid

Figure 10. Saponins and aromatic glycosides.

The compounds **1–76** were subjected to phytotoxicity assays to evaluate their effects on seed germination and the growth of roots and seedlings of various target organisms, including *Lactuca sativa*, *Solanum lycopersicum*, *Amaranthus retroflexus*, *Chenopodium album*, *Potamogeton oleracea*, and *Allium cepa* (Table 3). The assays were performed at different concentrations ranging from 10^{-4} to 10^{-9} M, following the protocol developed by Macias et al. [54], using the well-known herbicide Pendimethalin as a reference.

Table 3. Range of tested concentrations and target organisms for the phytotoxicity tests of the compounds **1–76**.

Compounds	Range of Concentrations	Organism Test					
		L. sativa	*S.lycopersicum*	*A. retroflexus*	*C. album*	*P. oleracea*	*A. cepa*
1–13	10^{-4}–10^{-7} M	x					
14–18	10^{-5}–10^{-7} M	x					
19–22	10^{-4}–10^{-7} M	x					
23–35	10^{-4}–10^{-8} M			x	x	x	
36–39	10^{-4}–10^{-7} M	x	x				
40–42	10^{-4}–10^{-8} M			x	x	x	
43–59	10^{-4}–10^{-9} M	x	x				x
60–62	10^{-4}–10^{-9} M	x	x				x
63–76	10^{-4}–10^{-8} M	x					

3.4.1. Assay with C_{13}-nor-Isoprenoids (**1–18**), Sesquiterpenes (**19–20**), Spirostane (**21**), and Pseudosapogenin (**22**)

Except for nor-terpenes **3**, **21**, and **22**, the tested compounds had no effect on germination but showed moderate inhibitory activity on root and shoot growth. The activity of the glycosylated compound **22** is intriguing, considering that glycosylation is the main detoxification mechanism adopted by plants to defend themselves against phytotoxic substances which they produce and store [55]. Among all the compounds tested, spirostane **21** was the most active, with root and shoot elongations reduced by up to 60% and germination by up to 30% at a concentration of 10^{-4} M. In general, the more polar compounds that are more soluble in water appear to be more active.

3.4.2. Assay with Lignans (**23–35**, **40–42**)

Compounds **23–35** and **40–42** were tested on *A. retroflexus*, *P. oleracea*, and *C. album*, in a concentration range varying from 10^{-4} to 10^{-8} M. Lignans **23–26** were the most active on *A. retroflexus*, inhibiting its germination even at the lowest concentration, while compounds **29–30** showed anti-germination activity on *P. oleracea* and anti-radical activity on *A. retroflexus*. All compounds were slightly stimulating for shoot elongation of *C. album* and *P. oleracea*.

3.4.3. Assay with Lignans (**36–39**)

These compounds were tested on *L. sativa* and *S. lycopersicum*, showing low phytotoxic activity on both. Of the four compounds in question, only compound **39** was able to inhibit the shoot length of *S. lycopersicum* by about 50% at a concentration of 10^{-4} M.

3.4.4. Assay with Aromatic Compounds (**43–59**) and Flavones (**60–62**)

The aqueous infusion of *C. parqui* leaves was tested on the germination, root length, and shoot length of *L. sativa*, *S. lycopersicum*, and *A. cepa* [39]. The interesting results obtained suggested dividing the entire extract into three fractions, two obtained via extraction with methylene chloride and ethyl acetate, while the third was the remaining aqueous part (Scheme 1). From the first organic fraction, compounds **48–50**, **52–53**, **55–56**, **60**, and **62** were isolated, while from the second fraction, compounds **43–47**, **51**, **54**, **57–58**, and **61** were isolated. Compounds **43–62** were tested on the same target organisms used for the phytotoxic evaluation of the aqueous extract, and some of them were far more active than the herbicides used as reference standards.

Only aromatic compounds **55** and **56** on *L. sativa* and chalcone **60** on *A. cepa* showed weak inhibitory effects on germination, while all others were practically inactive. Results on root elongation showed that some compounds, such as product **45**, could have a phytotoxic effect on *S. lycopersicum* but have a stimulating effect on *A. cepa*, or compound **44**, stimulating for *A. cepa* but inhibiting for *L. sativa*. Compounds **45** and **48** were able to inhibit the shoot length of *S. lycopersicum* and *A. cepa* by 66% and 60%, respectively, at a concentration of 1 nM [46].

3.4.5. Assay with Oxylipins (63–76)

The oxylipins were also tested on *L. sativa* seeds but in a narrower concentration range, specifically between 10^{-4} and 10^{-8} M [48]. It is not easy to rationalize the results of the phytotoxicity of these compounds. For example, at a concentration of 10^{-4} M, compounds **63–68** showed weak inhibitory action on germination, with values around 10%, and action on the elongation of the hypocotyl and root with inhibition values around 20%. However, at a concentration 100 times lower, only compounds **65** and **66** maintained weak inhibitory activity on root growth, while the corresponding alcohols **63** and **64**, or the corresponding methyl esters **67** and **68**, were even slightly stimulatory. Alternatively, compounds **69** and **71** stimulate germination and inhibit radical elongation, while compounds **70** and **72** inhibit germination and stimulate radical elongation. In general, it seems that the compounds present phytotoxicity values closely related to their degree of unsaturation, as for compounds **63–68**, **73**, and **74**, or phytotoxicity values dependent on the number of hydroxyl functions, as for compounds **75** and **76**.

It is interesting to note that oxylipins seem to play a crucial role in intra- and extracellular communication in vertebrates, fungi, and plants. In microorganisms, these metabolites are involved in the regulation of cell growth and differentiation, while in plants, their role in defense mechanisms based on apoptosis processes in response to infections caused by pathogens seems to be proven.

3.5. *Other Isolated Metabolites*

3.5.1. Kaurenic Glycosides (77–78) with Strychnine-Like Action

Two kaurenic glycosides named carbossiparquin (**77**) and parquin (**78**) have been isolated from the leaves of *C. parqui*, whose structures have been determined using NMR techniques and mass spectrometry [56] (Figure 9 and Table 4).

These compounds are structurally very similar, differing only in the presence of a second carboxylic function at carbon C 4 of the first compound. It is noteworthy that compounds **77** and **78** are quite like two toxins with strychnine-like action, namely carboxyatractyloside (A) and atractyloside (B), isolated from *Atractylis gummifera* [57]. In mice, carboxiparquin (**77**) has an LD_{50} value of 4.3 mg kg^{-1} and is over 50-times more toxic than crude extracts of *C. parqui* leaves. It is interesting to note that this toxin causes lesions in both the kidneys and the liver, like those observed in animals intoxicated after consuming *C. parqui*. The second compound (**78**) is relatively non-toxic and considered essentially a co-metabolite.

3.5.2. Cytotoxic Secondary Metabolites

Four new steroid saponins have also been isolated, three of which are monodesmosidic, called parquisoside A (**79**) and B (**80**) [58] and parquispiroside (**83**) [59], along with compound **84**, named parquifuroside [59]; together with the known steroid saponins: neotigogenin (**81**) [60] and (25R)-isonuatigenin (**82**) [61], capsicoside D (**85**) [62], 22-*O*-methyl-capsicoside D (**86**) [61], and digitogenin (**87**) [25,61]; the glycoalkaloid solasoninee (**88**) [62]; and the aromatic glycoside benzyl primeveroside (**89**) [59] (Figure 10 and Table 4). If compounds **79** and **80** are likely capable of inhibiting carrageenan-induced edema, there is no definitive evidence to support this. However, compounds **81–83** and **86–89** were tested for their cytotoxicity on four human cell lines: HeLa, HepG2, U87, and MCF7. Of these latter five compounds, only compound **81** showed moderate activity, with IC50 values of 7.7, 7.2, 14.1, and 3.3 μM, respectively. These values are quite promising considering that cisplatin, an antineoplastic chemotherapeutic agent used in the treatment of numerous tumors but with significant side effects, has much higher LC_{50} values of 39.2, 14.6, 7.3, and 23.0 μM, respectively [58].

Table 4. Other metabolites isolated from the leaves of *C. parqui*.

No.	Common Name/IUPAC Name	Ref.
77	Carboxiparquin	[56]
78	Parquin	
79	Parquisoside A/(3β,24S,25S)-spirost-5-ene-3,24-diol 3-O-{[α-L-rhamnopyranosyl-(1→2)]-β-D-glucopyranosyl-(1→4)-α-L-rhamnopyranosyl-(1→4)]-β-D-glucopyranoside	[58]
80	Parquisoside B/(3β,24S,25S)-spirost-5-ene-3,24-diol 3-O-{[α -L-rhamnopyranosyl-(1→4)-α-L-rhamnopyranosyl-(1→2)]-β-D-glucopyranosyl-(1→4)-α-L-rhamnopyranosyl-(1→4)}-β-D-glucopyranoside	[58]
81	Neotigogenin	[59]
82	(25R)-Isonuatigenin	[59]
83	Parquispiroside/25(R)-3β-[(O-β-D-glucopyranosyl-(1→3)-β-D-glucopyranosyl-(1→2)-O-[β-D-xylopyranosyl-(1→3)-O-β-D-glucopyranosyl-(1→4)-β-D-galactopyranosyl)oxy]-5α,15β,22R,25R-spirostan-3,15-diol	[58]
84	Parquifuroside/25(R)-26-[(β-D-Glucopyranosyl)oxy]-(3β[(O-β-D-glucopyranosyl-(1→3)-β-D-glucopyranosyl-(1→2)-O-[β-D-xylopyranosyl-(1→3)-O-β-D-glucopyranosyl-(1→4)-β-D-galactopyranosyl)oxy],5α,15β,22R,25R)-furostane-3,15,22-triol	[58]
85	Capsicoside D	[62]
86	22-O-Methylcapsicoside D	[58]
87	Digitogenin	[25,62]
88	Solasonine	[62]
89	Benzyl primeveroside	[58]
90	Ursolic acid	[58]
91	Oleanolic acid	[19,62]

4. Conclusions

Many Solanaceae are edible plants and are essential for human nutrition, but there are some which are extremely toxic, such as *Cestrum parqui* L'Herit, to the point that exposure to its leaves can cause respiratory difficulty, nausea, headache, and other unpleasant symptoms.

In the traditional medicine of some countries, *C. parqui* is used as an antipyretic and for the treatment of fever and inflammation. Its extracts showed moderate antimicrobial activity and possible anticancer and antiproliferative action on specific cell lines. Numerous studies have allowed for the isolation and the structural determination of just under one hundred secondary metabolites, such as C_{13}-nor-isoprenoids, sesquiterpenes, lignans, aromatic compounds, flavones, kaurenic glycosides, saponins, and alkaloids. Many of these compounds, but not all, have been studied to evaluate their insecticidal, antifeedant, and herbicidal activities both on weedy plants such as *A. retroflexus*, *P. oleracea*, and *C. album*, and on cultivated plants such as *A. cepa*, *L. sativa*, and *S. lycopersicum*.

In several cases, the activities measured proved to be higher. However, to date, the results are not definitive because it is not yet clear whether it is preferable to use an alcoholic or hydroalcoholic extract of the plant leaves or the individual metabolites. All of this suggests and justifies the significant interest in this plant, with possible and concrete commercial application.

Author Contributions: Conceptualization, M.C.D.M. and C.D.M.; resources, A.R.B. and S.P.; writing, P.N. and D.P.; supervision, A.D.M. All authors have read and agreed to the published version of the manuscript.

Funding: This research received no external funding.

Data Availability Statement: Not applicable.

Acknowledgments: This study was supported by AIPRAS Onlus (Associazione Italiana per la Promozione delle Ricerche sull'Ambiente e la Salute umana).

Conflicts of Interest: The authors declare no conflicts of interest.

References

1. Dewick, P.M. *Medicinal Natural Products: A Biosynthetic Approach*; John Wiley & Sons: Hoboken, NJ, USA, 2002.
2. Zhan, C.; Shen, S.; Yang, C.; Liu, Z.; Fernie, A.R.; Graham, I.A.; Luo, J. Plant metabolic gene clusters in the multi-omics era. *Trends Plant Sci.* **2022**, *27*, 981–1001. [CrossRef]
3. Souto, A.L.; Sylvestre, M.; Tölke, E.D.; Tavares, J.F.; Barbosa-Filho, J.M.; Cebrián-Torrejón, G. Plant-Derived Pesticides as an Alternative to Pest Management and Sustainable Agricultural Production: Prospects, Applications and Challenges. *Molecules* **2021**, *26*, 4835. [CrossRef]
4. Available online: https://keyserver.lucidcentral.org/weeds/data/media/Html/cestrum_parqui.htm (accessed on 5 May 2024).
5. L'Héritier de Brutelle, C.L.; Redouté, P.J.; Fréret, L. *Stirpes Novae aut Minus Cognitae, Quas Descriptionibus et Iconibus*; Ex Typographia Philippi-Dionysii Pierres; Nabu Press: Miami, NY, USA, 1784.
6. Ruiz, H. *Travels of Ruiz, Pavon, and Dombey in Peru and Chile*; Field Museum Press: Chicago, IL, USA, 1940.
7. Hunziker, A.T. *Genera Solanacearum: The Genera of the Solanaceae Illustrated, Arranged According to a New System*; Gantner Verlag: Ruggell, Liechtenstein, 2001.
8. Available online: https://www.actaplantarum.org/flora/flora_info.php?id=2055&nnn=Cestrum%20parqui%20L'H%C3%A9r (accessed on 17 July 2024).
9. Olmstead, R.G.; Bohs, L.; Migid, H.A.; Santiago-Valentin, E.; Garcia, V.F.; Collier, S.M. A molecular phylogeny of the Solanaceae. *Taxon* **2008**, *57*, 1159–1181. [CrossRef]
10. Olmstead, R.G. Phylogeny and Biogeography in Solanaceae, Verbenaceae and Bignoniaceae: A Comparison of Continental and Intercontinental Diversification Patterns. *Bot. J. Linn. Soc.* **2013**, *171*, 80–102. [CrossRef]
11. Available online: https://powo.science.kew.org/taxon/urn:lsid:ipni.org:names:30046363-2 (accessed on 10 May 2024).
12. Available online: https://europlusmed.org/cdm_dataportal/taxon/51664b74-b63f-4381-af29-46de5ef8d171 (accessed on 18 July 2024).
13. Available online: www.worldplants.de (accessed on 10 May 2024).
14. Groves, R.H.; Boden, R.; Lonsdale, W.M. *Jumping the Garden Fence: Invasive Garden Plants in Australia and Their Environmental and Agricultural Impacts*; WWF: Sydney, Australia, 2005.
15. Estomba, D.; Ladio, A.; Lozada, M. Plantas medicinales utilizadas por una comunidad Mapuche en las cercanías de Junín de los Andes, Neuquén. *BLACPMA* **2005**, *4*, 107–112.
16. Backhouse, N.; Delporte, C.; Negrete, R.; Salinas, P.; Pinto, A.; Aravena, S.; Cassels, B.K. Antiinflammatory and Antipyretic Activities of Cuscuta chilensis, *Cestrum parqui*, and Psoralea glandulosa. *Int. J. Pharmacogn. Pharm.* **1996**, *34*, 53–57. [CrossRef]
17. Molgaard, P.; Holler, J.G.; Asar, B.; Liberna, I.; Rosenbæk, L.B.; Jebjerg, C.P.; Jørgensen, L.; Lauritzen, J.; Guzman, A.; Adsersen, A.; et al. Antimicrobial evaluation of Huilliche plant medicine used to treat wounds. *J. Ethnopharmacol.* **2011**, *138*, 219–227. [CrossRef]
18. Chenni, H. Apoptosis induction by *Cestrum parqui* L'Hér. leaves on HL-60 Cell Line: Identification of active phytomolecules. *Int. J. Cancer. Stud. Res.* **2015**, *1*, 1–8.
19. Chenni, H. Identification of esterified oleanolic acid in *Cestrum parqui* leaves and its apoptotic induction on HT-29 cell line. *J. Med. Pharm. Innov.* **2015**, *2*, 63–68.
20. Falkenberg, S.S.; Tarnow, I.; Guzman, A.; Mølgaard, P.; Simonsen, H.T. Mapuche herbal medicine inhibits blood platelet aggregation. *Evid. Based Complement. Alternat. Med.* **2012**, *2012*, 647620. [CrossRef]
21. Souad, K.; Ali, S.; Mounir, A.; Mounir, T.M. Spermicidal activity of extract from *Cestrum parqui*. *Contraception* **2007**, *75*, 152–156. [CrossRef] [PubMed]
22. El-Sayed, A.S.A.; George, N.M.; Abou-Elnour, A.; El-Mekkawy, R.M.; El-Demerdash, M.M. Production and bioprocessing of camptothecin from *Aspergillus terreus*, an endophyte of *Cestrum parqui*, restoring their biosynthetic potency by *Citrus limonum* peel extracts. *Microb. Cell Fact.* **2023**, *22*, 4. [CrossRef] [PubMed]
23. Demain, A.L.; Vaishnav, P. Natural products for cancer chemotherapy. *Microb. Biotechnol.* **2011**, *4*, 687–699. [CrossRef] [PubMed]
24. Canham, P.A.S.; Warren, F.L. Saponins II. Isolation of gitogenin and digitogenin from *Cestrum parqui*. *S. Afr. J. Chem.* **1950**, *3*, 63–65.
25. Bianchi, E.; Girardi, F.; Diaz, F.; Sandoval, R.; Gonzales, M. Components of the leaves and fruit of *Cestrum parqui*: Tigogenin, digallogenin, digitogenin and ursolic acid. *Ann. Chim.-Rom.* **1963**, *53*, 1761–1778.

26. Bezerra, J.J.L.; Pinheiro, A.A.V.; de Lucena, R.B. Poisoning in ruminants caused by species of the genus *Cestrum* L. (Solanaceae) in Brazil: A review of toxicological and phytochemical evidence. *Toxicon* **2023**, *236*, 107348. [CrossRef] [PubMed]
27. Brevis, C.; Quezada, M.; Sierra, M.A.; Carrasco, L.; Ruiz, A. Lesiones observadas en intoxicaciones accidentales con *Cestrum parqui* (L'Herit) en bovinos. *Arch.Med. Vet.* **1999**, *31*, 109–118. [CrossRef]
28. Barbouche, N.; Hajem, B.; Lognay, G.; Ammar, M. Contribution for studying the biological activity of the leaves extracts of *Cestrum parqui* L'Herit. on desert locust *Schistocera gregaria*. *Biotechnol. Agron. Soc.* **2001**, *5*, 85–90.
29. Ikbal, C.; Ben, H.K.; Ben, H.M. Insect growth regulator activity of *Cestrum parqui* saponins: An interaction with cholesterol metabolism. *Commun. Agric. Appl. Biol. Sci.* **2006**, *71*, 489–496.
30. Zapata, N.; Budia, F.; Viñuela, E.; Medina, P. Insecticidal effects of various concentrations of selected extractions of *Cestrum parqui* on adult and immature Ceratitis capitata. *J. Econ. Entomol.* **2006**, *99*, 359–365. [CrossRef] [PubMed]
31. Thomas, M.C.; Heppner, J.B.; Woodruff, R.E.; Weems, H.V.; Steck, G.J.; Fasulo, T.R. *Mediterranean Fruit Fly, Ceratitis Capitata. (Wiedemann) (Insecta: Diptera, Tephritidae)*; University of Florida: Gainesville, FL, USA, 2001; IFASA Extension 2001, EENY-214.
32. Huerta, A.; Chiffelle, I.; Araya, L.; Curkovic, T.; Araya, J.E. Insecticidal capacity of *Cestrum parqui* (Solanaceae) leaf extracts on *Xanthogaleruca luteola* (Coleoptera: Chrysomelidae) adults. *Rev. Colomb. Entomol.* **2021**, *47*, e10492. [CrossRef]
33. Ben Hamouda, A.; Ammar, M.; Habib, M.; Hamouda, B. Effect of Olea europea and *Cestrum parquii* Leaves on the Cuticle and Brain of the Desert Locust, *Schistocerca gregaria* Forsk (Orthoptera:Acrididae). *Pest Technol.* **2011**, *5*, 55–58.
34. Chaieb, I.; Halima-Kamel, M.B.; Ben, M.H. Antifeedant activity of *Cestrum parqui* crude saponic extract. *Tunis. J. Med. Plants Nat. Prod.* **2009**, *1*, 27–33.
35. Chaieb, I.; Mounia, B.H.; Habib, B.H. Toxicity experiments of the saponic extract of *Cestrum parqui* (Solanaceae) on some insect spices. *J. Entomol.* **2007**, *4*, 113–120.
36. Chaieb, I.; Mounia, B.H. The effect of food containing *Cestrum paquii* (Solanaceae) extract on various damaging Lepidoptera. *Meded. Rijksuniv. Gent. Fak. Landbouwkd. Toegepaste Biol. Wet.* **2001**, *66*, 479–490.
37. Chaieb, I.; Habib, B.; Hichem, B.J.; Monia, B.H.; Habib, B.H.; Zine, M. Purification of a natural insecticidal substance from *Cestrum parqui* (Solanaceae). *PJBS* **2007**, *10*, 3822–3828.
38. Chaieb, I.; Ben, H.; Trabelsi, M.; Hlawa, W.; Raouani, N.; Ben, A.; Daami, M.; Ben, H. Pesticidal Potentialities of *Cestrum parqui* saponins. *Int. J. Agric. Res.* **2003**, *2*, 275–281.
39. Huanquilef, C.; Espinoza, J.; Mutis, A.; Bardehle, L.; Hormazábal, E.; Urzúa, A.; Quiroz, A. Antifeedant activities of organic fractions from *Cestrum parqui* leaves on the red-haired bark beetle *Hylurgus ligniperda*. *J. Soil Sci. Plant Nutrit.* **2021**, *21*, 13–21. [CrossRef]
40. Chaieb, I.; Halima-Kamel, M.B.; Hammouda, M.B. Experiments for studying the molluscicidal potential of *Cestrum parqui* (Solanaceae) saponins against *Theba pisana* (Helicidae) snails. *Comm. Agric. Appl. Biol. Sci.* **2005**, *70*, 809–815.
41. Chaieb, I.; Tayeb, W. Comparison of the molluscicidal activity of *Cestrum parqui* (Solanaceae) and *Quillaja saponaria* (Quillajaceae) saponins. *Tunis. J. Med. Plants Nat. Prod.* **2009**, *2*, 31–35.
42. D'Abrosca, B.; DellaGreca, M.; Fiorentino, A.; Monaco, P.; Oriano, P.; Temussi, F. Structure elucidation and phytotoxicity of C_{13}-nor-isoprenoids from *Cestrum parqui*. *Phytochemistry* **2004**, *65*, 497–505. [CrossRef]
43. D'Abrosca, B.; DellaGreca, M.; Fiorentino, A.; Monaco, P.; Natale, A.; Oriano, P.; Zarrelli, A. Structural characterization of phytotoxic terpenoids from *Cestrum parqui*. *Phytochemistry* **2005**, *66*, 2681–2688. [CrossRef] [PubMed]
44. Fiorentino, A.; DellaGreca, M.; D'Abrosca, B.; Oriano, P.; Golino, A.; Izzo, A.; Monaco, P. Lignans, neolignans and sesquilignans from *Cestrum parqui* l'Her. *Biochem. Syst. Ecol.* **2007**, *35*, 392–396. [CrossRef]
45. D'Abrosca, B.; DellaGreca, M.; Fiorentino, A.; Golino, A.; Monaco, P.; Zarrelli, A. Isolation and characterization of new lignans from the leaves of *Cestrum parqui*. *Nat. Prod. Res.* **2006**, *20*, 293–298. [CrossRef]
46. D'Abrosca, B.; DellaGreca, M.; Fiorentino, A.; Monaco, P.; Zarrelli, A. Low molecular weight phenols from the bioactive aqueous fraction of *Cestrum parqui*. *J. Agric. Food Chem.* **2004**, *52*, 4101–4108. [CrossRef] [PubMed]
47. DellaGreca, M.; Fiorentino, A.; Izzo, A.; Napoli, F.; Purcaro, R.; Zarrelli, A. Phytotoxicity of secondary metabolites from *Aptenia cordifolia*. *Chem. Biodivers.* **2007**, *4*, 118–128. [CrossRef] [PubMed]
48. Fiorentino, A.; D'Abrosca, B.; DellaGreca, M.; Izzo, A.; Natale, A.; Pascarella, M.T.; Pacifico, S.; Zarrelli, A.; Monaco, P. Chemical characterization of new oxylipins from *Cestrum parqui*, and their effects on seed germination and early seedling growth. *Chem. Biodivers.* **2008**, *5*, 1780–1791. [CrossRef] [PubMed]
49. Cangiano, T.; DellaGreca, M.; Fiorentino, A.; Isidori, M.; Monaco, P.; Zarrelli, A. Lactone diterpenes from the aquatic plant *Potamogeton natans*. *Phytochemistry* **2001**, *56*, 469–473. [CrossRef] [PubMed]
50. Cangiano, T.; DellaGreca, M.; Fiorentino, A.; Isidori, M.; Monaco, P. Effect of ent-labdane diterpenes from Potamogetonaceae on *Selenastrum capricornutum* and other aquatic organisms. *J. Chem. Ecol.* **2002**, *28*, 1091–1102. [CrossRef] [PubMed]
51. Pollio, A.; Romanucci, V.; Di Mauro, A.; Barra, F.; Pinto, G.; Crescenzi, E.; Roscetto, E.; Palumbo, G. Polyphenolic profile and targeted bioactivity of methanolic extracts from Mediterranean ethnomedicinal plants on human cancer cell lines. *Molecules* **2016**, *21*, 395. [CrossRef] [PubMed]
52. Cutillo, F.; D'Abrosca, B.; DellaGreca, M.; Zarrelli, A. Chenoalbicin, a novel cinnamic acid amide alkaloid from *Chenopodium album*. *Chem. Biodivers.* **2004**, *1*, 1579–1583. [CrossRef] [PubMed]
53. DellaGreca, M.; Previtera, L.; Purcaro, R.; Zarrelli, A. Cinnamic ester derivatives from Oxalis pes-caprae (*Bermuda buttercup*). *J. Nat. Prod.* **2007**, *70*, 1664–1667. [CrossRef] [PubMed]

54. Macıas, F.A.; Castellano, D.; Molinillo, J.M.G. Search for standard phytotoxicity bioassay for allelochemicals. Selection of standard target species. *J. Agric. Food Chem.* **2000**, *48*, 2512–2521. [CrossRef] [PubMed]
55. le Roy, J.; Huss, B.; Creach, A.; Hawkins, S.; Neutelings, G. Glycosylation is a major regulator of phenylpropanoid availability and biological activity in plants. *Front. Plant Sci.* **2016**, *7*, 735. [CrossRef] [PubMed]
56. Pearce, C.M.; Skelton, N.J.; Naylor, S.; Kanaan, R.; Kelland, J.; Oelrichs, P.B.; Sanders, J.K.M.; Williams, D.H. Parquin and carboxyparquin, toxic kaurene glycosides from the shrub *Cestrum parqui*. *J. Chem. Soc.* **1992**, *1*, 593–600. [CrossRef]
57. Daniele, C.; Dahamna, S.; Firuzi, O.; Sekfali, N.; Saso, L.; Mazzanti, G. *Atractylis gummifera* L. poisoning: An ethnopharmacological review. *J. Ethnopharmacol.* **2005**, *97*, 175–181. [CrossRef] [PubMed]
58. Baqai, F.T.; Ali, A.; Ahmad, V. Two new spirostanol glycosides from *Cestrum parqui*. *Helv. Chim. Acta* **2001**, *84*, 3350–3356. [CrossRef]
59. Mosad, R.R.; Ali, M.H.; Ibrahim, M.T.; Shaaban, H.M.; Emara, M.; Wahba, A.E. New cytotoxic steroidal saponins from *Cestrum parqui*. *Phytochem. Lett.* **2017**, *22*, 167–173. [CrossRef]
60. Abdel-Gwad, M.M.; El-Amin, S.M.; El-Sayed, M.M.; Refahy, L.A.; Sabry, W.A. Molluscicidal saponins from *Cestrum parqui*. *AJPS* **1997**, *20*, 80–84.
61. Torres, R.; Modak, B.; Faini, F. (25R)-Isonuatigenin, an unusual steroidal sapogenin as taxonomicmarker in *Cestrum parqui* and Vestia lycioides. *Biol. Soc. Chil. Quim.* **1988**, *33*, 239–241.
62. Silva, M.; Mancinelli, P.; Cheul, M. Chemical study of *Cestrum parqui*. *J. Pharm. Sci.* **1962**, *51*, 289. [CrossRef] [PubMed]

plants

MDPI

Review

The Lack of Standardization and Pharmacological Effect Limits the Potential Clinical Usefulness of Phytosterols in Benign Prostatic Hyperplasia

Mădălina-Georgiana Buț [1,2], George Jîtcă [3,*], Silvia Imre [4], Camil Eugen Vari [3], Bianca Eugenia Ősz [3], Carmen-Maria Jîtcă [1]

1 Doctoral School of Medicine and Pharmacy, I.O.S.U.D, George Emil Palade University of Medicine, Pharmacy, Science and Technology of Târgu Mures, 540139 Târgu Mures, Romania; madalina-georgiana.batrinu@umfst.ro (M.-G.B.); carmenrusz20@gmail.com (C.-M.J.)

2 Department of Biochemistry, Faculty of Pharmacy, George Emil Palade University of Medicine, Pharmacy, Science and Technology of Târgu Mures, 540139 Târgu Mures, Romania; amelia.tero-vescan@umfst.ro

3 Department of Pharmacology and Clinical Pharmacy, Faculty of Pharmacy, George Emil Palade University of Medicine, Pharmacy, Science and Technology of Târgu Mures, 540139 Târgu Mures, Romania; camil.vari@umfst.ro (C.E.V.); bianca.osz@umfst.ro (B.E.Ő.)

4 Department of Analytical Chemistry and Drug Analysis, Faculty of Pharmacy, George Emil Palade University of Medicine, Pharmacy, Science and Technology of Târgu Mures, 540139 Târgu Mures, Romania; silvia.imre@umfst.ro

* Correspondence: george.jitca@umfst.ro

Abstract: The prevalence of benign prostatic hyperplasia (BPH) markedly increases with age. Phytotherapeutic approaches have been developed over time owing to the adverse side effects of conventional medications such as 5-reductase inhibitors and α1-adrenergic receptor antagonists. Therefore, dietary supplements (DS) containing active compounds that benefit BPH are widely available. Phytosterols (PSs) are well recognized for their role in maintaining blood cholesterol levels; however, their potential in BPH treatment remains unexplored. This review aims to provide a general overview of the available data regarding the clinical evidence and a good understanding of the detailed pharmacological roles of PSs-induced activities at a molecular level in BPH. Furthermore, we will explore the authenticity of PSs content in DS used by patients with BPH compared to the current legislation and appropriate analytical methods for tracking DS containing PSs. The results showed that PSs might be a useful pharmacological treatment option for men with mild to moderate BPH, but the lack of standardized extracts linked with the regulation of DS containing PSs and experimental evidence to elucidate the mechanisms of action limit the use of PSs in BPH. Moreover, the results suggest multiple research directions in this field.

Keywords: phytosterol; benign prostatic hyperplasia; sitosterol; campesterol; dietary supplement

1. Introduction

Benign prostatic hyperplasia (BPH) is a common and progressive condition that affects the quality of life of men in an age-dependent manner, being present in 15–60% of men over 40 years [1–3]. BPH is a non-malignant enlargement of the prostate that can lead to obstruction and irritation of the lower urinary tract. As the prostate gland increases in volume, constriction of the urethra occurs with the appearance of symptoms such as weak urinary flow, incomplete emptying of the bladder, nocturia, or dysuria. These symptoms are associated with BPH and are referred to as lower urinary tract symptoms (LUTS) [4]. The etiology of BPH is influenced by a wide variety of risk factors, such as age, hormonal imbalance, inflammation, metabolic syndrome, oxidative stress, or inhibition of apoptosis in prostate tissue [4].

Reducing LUTS and improving the quality of life (QL) are the primary goals of BPH treatment. As therapy has changed significantly over the last decade, the number of surgeries has steadily decreased while the number of cases treated with medicine has increased [5,6]. Currently, two large classes of drugs are used in pharmaceutical practice for the treatment of BPH: α1-adrenergic receptor antagonists (doxazosin, terazosin, and tamsulosin), which relieve LUTS by relaxing the smooth muscle of the prostate stroma and bladder neck, and 5α-reductase inhibitors (5-αRi) such as finasteride or dutasteride [7,8]. 5α-reductase (5-αR) is an enzyme of major importance in the development of BPH and is responsible for the formation of dihydrotestosterone (DHT), the main biologically active metabolite of testosterone (T). It is well known that androgen excess, mainly DHT, has been suggested to be associated with the development of BPH [9,10]. However, these prescription-only medicines are used in advanced forms of BPH, according to the attending physician's recommendation. Although there is a significant clinical benefit when administered to BPH patients [11,12], conventional therapy is correlated with disorders of sexual dynamics in men, such as erectile dysfunction, increased risk of impotence, ejaculation disorders, gynecomastia, and orthostatic hypotension [11,13–15].

Growing interest in the use of dietary supplements (DS) for healthcare management has led to the availability of natural products in the market. Consequently, an increasing number of patients tend to adopt a plant-based treatment for BPH. Recent data present a wide range of bio-compounds, such as phytosterols, phenolic compounds (polyphenols, catechins), and fatty acids, which can be further formulated as adjuvant medication to conventional therapy. For instance, phytosterols (sitosterol, stigmasterol, campesterol), which have a similar chemical structure to synthetic 5-αRi, represent the most promising class derived from plants with inhibitory action on 5-αR. Phytosterols (PSs) are increasingly used to alleviate BPH symptoms, and DS manufacturers promote their beneficial effects without substantial experimental evidence. The National Institutes of Health (NIH) Dietary Supplement Label Database (DSLD) releases a database with info on over 5000 dietary supplement labels for supporting healthy prostate function in men [16]. The most common sources of PSs found in DS are *Serenoa repens* (54%), *Pygeum africanum* (15%), *Urtica dioica* (14%), *Curcupita pepo* (14%), or sitosterol in the singular form [17]. Available published data focus mainly on establishing and demonstrating the beneficial effects of plant-based DS in BPH [3,18,19]; however, the underlying mechanisms of PSs involved have not been thoroughly investigated.

As over-the-counter medications, PS-based DS are available in pharmacies, natural stores, or online and information accessible to users is skewed toward commercial interests, making them vulnerable to hyperbolic advertising and misleading claims. Since the regulatory aspects of the DS industry are ambiguous and incomplete, manufacturers set their own standards. Our understanding of plant extracts is limited because of the lack of standardization. Thus, clinical trials are difficult to interpret, and the therapeutic efficacy of phytocompounds is inconsistent, resulting in a lack of progress in clinical practice and public health [20]. In this context, another difficulty is brought about by challenges in the analytical field, and the lack of standardized analysis techniques for monitoring the quality of PSs in DS is absolutely necessary.

In light of the growing BPH burden around the world, the aim of this study is to investigate the role of PSs in BPH. We sought to provide the limitations and difficulties associated with using PSs for BPH, considering the potential therapeutic benefits of PSs, their low pharmaco-toxicological profile, the difficulty in assessing their clinical effectiveness, as well as the growing interest in the consumption of PSs. Thus, the purpose of this paper is to review the molecular-level mechanisms of action of PSs intervening in BPH and the clinical evidence, the quality and quantity of PSs used by patients with BPH compared to the legislation in force, and the appropriate analytical methods reported in the literature used for qualitative and quantitative monitoring of DS with PSs content.

To the best of our knowledge, this is the first study to assess, in particular, the effects of phytosterols from multiple views in BPH. To fill a gap in the existing literature and uncover the nature of review novelty, critical building blocks include chemical properties, sources of PSs, analytical methods for tracking DS containing PSs, regulatory framework, molecular mechanisms, and clinical evidence for PSs in BPH.

2. Phytosterols—Characteristics

2.1. Chemical Properties of Plant Sterols

Structurally, PSs are bioactive compounds and plant equivalents of cholesterol that are specific to animal groups. Steroidal in shape, they contain a hydroxyl group at C_3, a hydrocarbonate chain at C_{17}, and one or more double bonds at C_5 of the base skeleton [21–23], see Figures 1 and 2. Phytostanols are a subgroup of this class, with a fully saturated basic skeleton [23,24].

Despite the presence of over 200 PSs, β-sitosterol (SIT), stigmasterol, and campesterol, are the essential phytosterols to consider. As the dominant compound, SIT accounted for 90% [23]. These data were confirmed by NIH-DSLD, which released label statements with DS containing mostly SIT, campesterol, and stigmasterol. However, most previous studies have focused on the same phytosterol for various effects, such as anti-obesity, anti-diabetic, anti-microbial, anti-inflammatory, immunomodulatory, and anti-cancer effects [16].

PSs can be found in a free form (FPS) or conjugated form (CPS) at the hydroxyl group from position 3 (Figure 1). The four common types of conjugate sterol lipid classes are sitosteryl fatty acid ester (SFE) which is the most common, sitosteryl glycoside (SG), acylated, sitosteryl glycoside (ASG), or hydroxycinnamate sitosteryl ester (HSE) [25,26] (Figure 2). There is no evidence that CPS has the same therapeutic effects as FPS in BPH, with CPS being less studied due to the lack of analytical reference standards. Therefore, PSs analysis should consider the form of the PSs present in the matrix (FPS or CPS) and use a suitable extraction method to ensure the determination of the total amount of PSs. Instead, CPS has a lower melting point and higher solubility in vegetable oils [27] and is used to enrich foods, such as salad dressings [28], margarine [29], or cheddar [30].

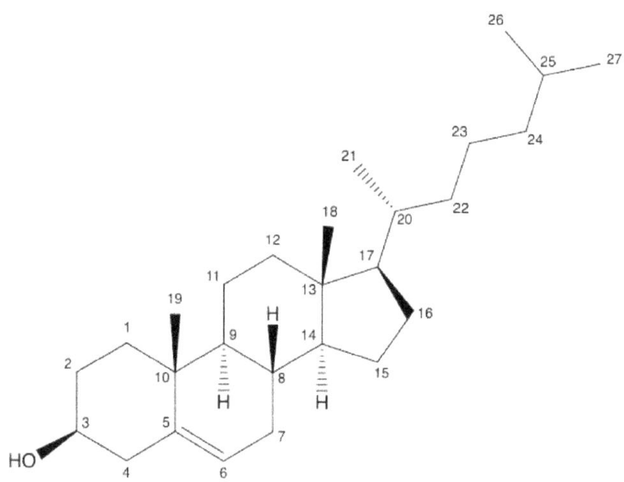

Figure 1. Cholesterol with carbon numbering according to IUPAC.

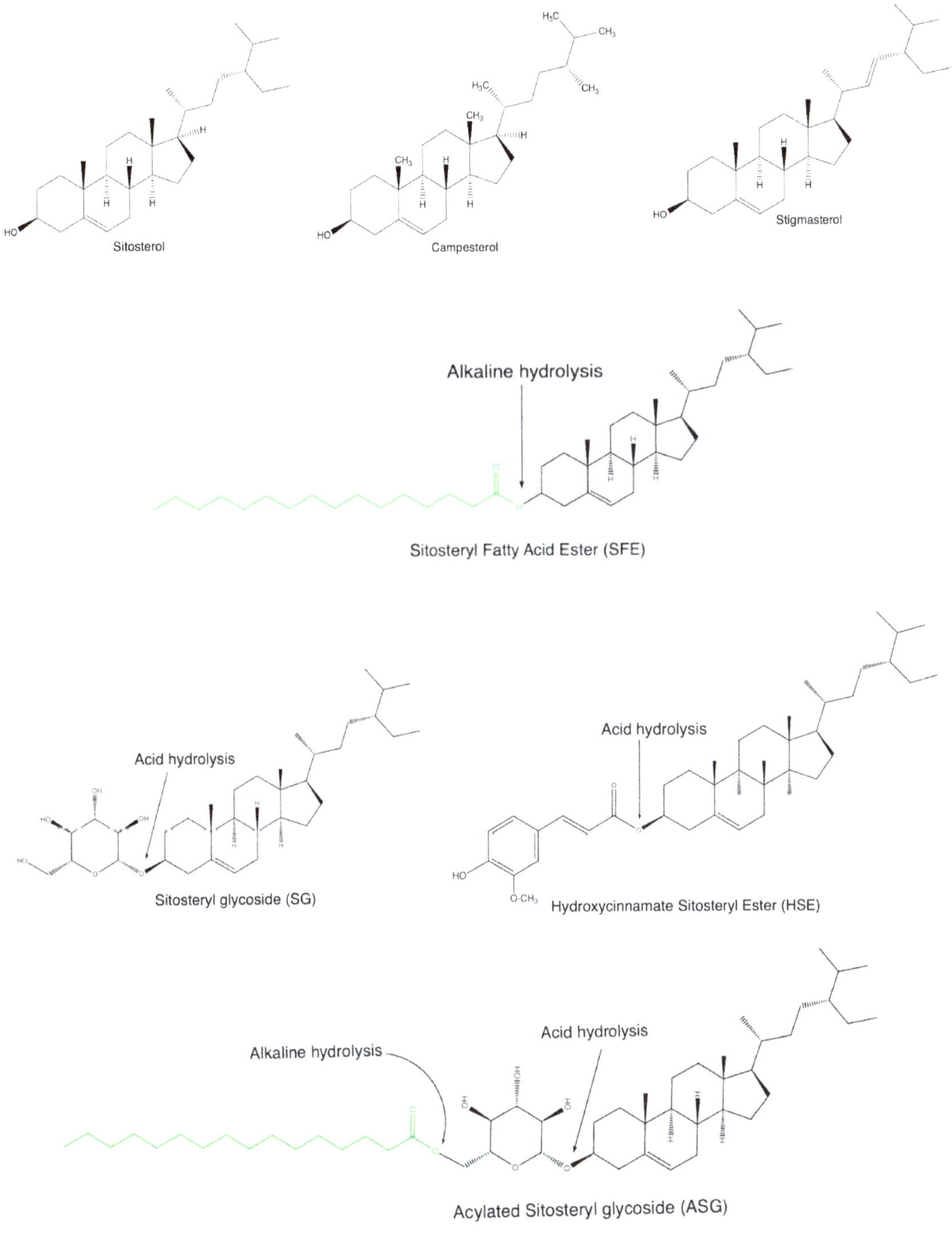

Figure 2. The structures of the most common phytosterols found in dietary supplements (free-form FPS and conjugated CPS); green radical—fatty acid residue.

2.2. Sources of Plant Sterols

PSs are ubiquitous compounds distributed in nature and are present in small amounts in daily food. As dietary sources, fruits and vegetables usually contain only small amounts of PSs (less than 0.05% on a wet basis) [25,26]; however, nuts and vegetable oils may contain more than 1% PSs [31]. On the other hand, phytostanols are present in cereals (corn, wheat, rye, rice), fruits, and vegetables, but their concentrations are generally lower than unsaturated plant sterols [32]. The typical daily consumption of PSs is between 140–400 mg/day in different populations depending on the country and type of diet, and it comes mostly from vegetable oils, cereals, fruits, and vegetables [33]. Average PSs ratios between 140–360 mg/day have been estimated in Finland; for a review, see [23,34], 163 mg/day United Kingdom [23], 100.6 mg/zi Brazil [35], or 392.3 mg/day in China [36]. Vegetarians, generally vegans, have the highest PSs intake of >1 g/day [37]. In DS, the main source of PSs is represented by standardized plant extracts, either as a single compound or as a combination of active principles. Dwarf palm extract (*Serenoa repens*) is the most commonly used DS formulation [38]. Consequently, the National Institute of Standards and Technology (NIST) has developed two standard reference materials (SRM), SRM 3250 *Serenoa fruits* and SRM 3251 *Serenoa repens*, which can be used as a reference for PSs monitoring in DS with *Serenoa repens* extract [39]. In addition to dwarf palm extract, other plant sources containing PSs used in BPH are *Curcubita pepo* seeds, *Epilobium parviflorum*, *Epilobium angustifolium* aerial parts, *Hypoxis hemerocallidea*, *Hypoxis rooperi* corn, *Prunus africana* bark, *Secale cereale* pollen, *Urtica dioica* root, and *Cernilton* [3,19].

3. Phytosterols—Beneficial Effects in BPH

3.1. Pathophysiology of Benign Prostatic Hyperplasia

Solid and complex pathophysiological backgrounds are involved in BPH. BPH is characterized by the proliferation of both stromal and epithelial elements, leading to gland enlargement and, in some cases, urinary obstruction [40]. Although the cause of BPH remains incompletely elucidated, it is known that a central role is played by the growth of both stromal and epithelial elements caused by excess androgens [41]. BPH does not occur in males castrated before puberty or in those with genetic diseases that block the activity of androgens [42]. The action of DHT is mediated via androgen receptors (ARs), a ligand-dependent nuclear transcription factor and a member of the steroid hormone nuclear receptor family, which regulates the expression of genes that support the growth and survival of prostatic epithelial and stromal cells. Although T can also bind to ARs and stimulate growth, DHT is ten times more potent [43].

In addition, the male body stimulates the conversion of T to estrogen (ES) by increasing aromatase activity due to the high percentage of adipose tissue associated with advancing age. The circulating ES/T ratio increases, resulting in a decreased negative androgen control of ES release [4,44]. Estrogen also plays an important role in the development of BPH, and its level is correlated with prostate volume [45]. In contrast, the study conducted by Miwa et al. did not identify the same relationship between ES levels and prostate volume [46]. Estrogen receptor (ER) types are thought to have a major influence on the effects of ES on prostate tissues. Specifically, ER-α stimulation causes hyperplasia, dysplasia, and inflammation [47]. Conversely, ER-β activation decreased proliferation and promoted apoptosis in BPH in an androgen-independent manner [48]. Therefore, ER-α/ER-β ratio plays an important role in ES-induced cell proliferation.

Several previous experimental studies have linked chronic inflammation with the development of BPH, suggesting that chronic inflammation may contribute to the development of the disease [49–51]. The role of inflammation in the development of BPH is underlined by the strong correlation between inflammation proven by histological criteria, the International Prostate Symptom Score (IPSS), and prostate volume; thus, the inflammatory process is considered a therapeutic target in BPH [49]. A study conducted by Nickel et al. analyzed prostate tissue from 374 patients who underwent transurethral resection of the prostate (TURP) for BPH and noted the presence of chronic or acute inflammation in

70% of patients [52]. Cyclooxygenase-2 (COX-2) expression is associated with inflammatory processes in BPH. Prostaglandins are a group of pro-inflammatory mediators, synthesized from arachidonic acid, under the action of COX-1 and COX-2, identified in the prostate tissue of BPH patients [53,54]. Furthermore, a study by Wang and his colleagues suggested that up-regulation of anti-apoptotic proteins correlated with increased COX-2 expression inhibits prostate apoptosis in BPH [55]. Other studies have demonstrated that the administration of COX-2 inhibitors to patients with BPH produces a significant increase in apoptotic processes [56]. Additionally, another study has demonstrated that induced nitric oxide synthase (iNOS) is only present in patients with BPH and contributes to inflammation [57].

Oxidative stress is another trigger that can lead to BPH development and progression. A study using an animal model of BPH reported decreased activity of antioxidant systems such as glutathione, superoxide dismutase, glutathione peroxidase, and catalase. Furthermore, a significant increase in the lipid peroxidation process in BPH has been reported, which was inhibited by finasteride administration [58]. Considering the pathophysiological implications, the use of PSs in BPH is explained by their anti-androgenic, anti-inflammatory, and antioxidant effects alongside their ability to modulate apoptotic processes.

3.2. Molecular Mechanism of PSs in BPH Development

- Anti-androgenic effect of phytosterols in BPH

Plant-derived PSs that are structurally similar to synthetic 5-αRi (finasteride, dutasteride) represent a potentially highly effective class of 5-αRi. According to an in vitro study performed by Marisa Cabesa and her colleagues on hamster prostate tissue, SIT inhibits 5-αR in a dose-dependent manner (IC_{50} = 1.1 μg/mL compared to finasteride with IC_{50} = 0.003 μg/mL) [59]. Other data have shown that stigmasterol, extracted from *Phyllanthus urinaria*, also inhibits 5-αR and is less potent than SIT (IC_{50} = 11.2 μg/mL) [60]. To offset the low efficacy of PSs, a combination of conventional treatments with PSs may provide some clinical benefit. Studies have also been performed on plant extracts containing PSs with inhibitory action on 5-αR, without the individual identification and specific contribution of each phytocompound to 5-αR activity [61–63]. The results are summarized in Table 1. Among the extracts containing PSs, saw palmetto extract has been the most studied. An in vitro study performed on *Serenoa repens* extract using a baculovirus-directed insect cell expression system demonstrated its ability to inhibit 5-αR in a non-competitive manner [64]. Various saw palmetto extracts (hexane, ethanol, and hypercritical CO_2) have reported IC_{50} values between 25 μg/mL and 2200 μg/mL, depending on the modality of 5-αR activity assessment [65]. The data collected indicate that SIT, a standardized extract, is more potent than non-standardized extracts from plants, despite having a potency over 1000× lower than finasteride. Moreover, the extraction method used may affect the inhibitory activity of phytocompounds from plant products. A study conducted by Nahata et al. analyzed the inhibitory capacity of 5-αR on different extracts from *Urtica dioica* root (ethanolic, petroleum ether, and aqueous); the ethanolic extract was the most potent [66].

Table 1. Dietary components with 5-αR inhibitory effect.

Molecules	Sample	IC_{50} μg/mL	Ref.
Sitosterol Finasteride	standardized extract standard	1.1 0.003	[59]
Stigmasterol	*Phyllanthus urinaria*, extract from the whole plant	11.2	[60]
	Plant extract with PSs content		
Serenoa repens	whole plant	25–2200	[65]
Epilobium parviflorum	stem and leaves	160	[61]
Pygeum africanum	standardized extract	780	[62]
Curcubita pepo	pupmkin seed oil and soft extract	5880	[63]
Urtica dioica	standardized extract	14,700	[62]

In addition to the 5-αR inhibitory activity, PSs resulted in decreased AR expression and inhibition of the DHT-AR complex [67]. The estrogenic effects of PSs have recently been reviewed by Nattagh-Eshtvani E. Furthermore, colleagues [68]. According to an in vivo study, PSs have direct estrogenic activity and act as selective modulators of ES (SERMs) on both ER-α and ER-β. Although PSs interact with the ER, the administration of SIT in an animal model did not increase uterine weight, a key marker of estrogenic activity; for a review, see [68–70]. However, in BPH, there are no data proving the beneficial or non-beneficial effects of PSs through an estrogen-mediated mechanism; the estrogenic effect at the level of prostate tissue depends on the ability of PSs to stimulate ER but also on the ER-α/ER-β ratio.

- Phytosterols as anti-inflammatory and antioxidant dietary components in BPH

The anti-inflammatory effects of PSs have been studied and demonstrated in animals and humans under various pathological conditions. In a study performed in rats with non-alcoholic fatty liver disease, plant sterol-fortified skimmed milk was administered. After 12 weeks of treatment, inhibition of interleukin 6 (IL-6), interleukin 10 (IL-10), and C-reactive protein and improvement of hepatic steatosis were observed [71]. Researchers have found that treatment with SIT (20 mg/kg) for 8 weeks inhibited the activation of nuclear factor kappa B (NF-kB) and the production of pro-inflammatory cytokines in mice with high-fat-diet-induced intestinal injury and inflammation [72].

It is well known that NF-kB stimulates the expression of proteins that contribute to the pathogenesis of inflammation. NF-kB activation is a hallmark of inflammation [73]. Inhibition of NF-kB activation by PSs has been reported by several authors [74–76]. Moreover, recent data indicate that PSs alleviate the inflammatory reaction in lipopolysaccharides (LPS)-induced macrophage models and cell phagocytosis and inhibit the release of tumor necrosis factor-α (TNF-α), the expression and activity of COX-2, iNOS, and phosphorylated extracellular signal-regulated protein kinase (p-ERK). The anti-inflammatory activity of SIT was higher compared to that of stigmasterol and campesterol, suggesting that PSs without a double bond at C_{22} and with an ethyl group at C_{24} are more potent anti-inflammatory agents [77].

Based on the relationship between inflammation and BPH, an in vivo study tested the anti-inflammatory properties of *Serenoa repens* extracts enriched with 3% SIT (0.2–0.3% normally). The results showed that SIT significantly decreased the expression of COX-2 compared to the untreated BPH group [78]. In the same study, the group with BPH without treatment showed increased levels of NF-kB compared to the placebo group, and in the case of the group treated with *Serenoa repens* extract enriched with SIT, the gene expression of NF-kB was significantly inhibited [78]. Therefore, the anti-inflammatory effect of SIT in BPH is due to the inhibition of COX-2 and NF-κB expression. No studies have strictly evaluated the anti-inflammatory effects of PSs in BPH subjects.

Its anti-inflammatory effect is closely related to its antioxidant effects. Inflammation is one factor that produces reactive oxygen species (ROS) in the prostate tissue and is associated with oxidative stress [79], and an excess of ROS can also trigger the inflammatory process [80,81]. PSs have an antioxidant effect and act as free radical scavengers, cell membrane stabilizers, and antioxidant enzyme boosters; for a review, see [82,83]. However, in BPH, the antioxidant effects of PSs have not yet been tested. At the hepatic level, SIT attenuates alcohol-induced ROS by restoring erythrocyte membrane fluidity, reducing glutathione depletion malondialdehyde overproduction, restoring antioxidant enzyme activity, and reducing malondialdehyde overproduction [84].

- Apoptotic effect of phytosterols in BPH

Cell growth is controlled by a constant balance between the stimulation and inhibition of apoptosis-related metabolic pathways. In particular, the Bcl-2 family of proteins (B-cell lymphoma 2) and the Bcl-2 associated with protein X (BaX) play an essential role by modulating the activity of certain caspases, especially caspase-9. Bax-expressing cells undergo apoptosis, while Bcl-2-expressing cells undergo carcinogenesis, which results from suppressing apoptosis [85]. Specifically, increased levels of Bcl-2 inhibit Bax and prevent cytochrome C from being released into mitochondria, which inhibits the formation of necessary complexes of apoptotic protease activating factor-1 (APAF1), cytochrome C, and caspase-9, essential for cell survivor [86]. The Bax/Bcl-2 ratio has been suggested to play an important role in the regulation of apoptosis by androgens. Thus, androgen deficiency during apoptosis led to an increase in the Bax/Bcl-2 ratio [87]. Several studies have suggested that apoptosis is diminished in BPH. For example, a study conducted by Kyprianol et al. showed increased expression of Bcl-2 in epithelial cells compared to that in healthy prostate tissue [88]. Furthermore, other studies have shown increased levels of apoptotic inhibitory proteins in the human prostate with BPH [89,90].

Another pathway involved in the regulation of apoptosis is mitogen-activated protein kinase (MAPK). Three MAPK pathways are involved in cell cycle modulation: extracellular signal-regulated protein kinase (ERK), p38, and Jun N-terminal Kinase (JNK) [91]. JNK and p38 synergistically promote apoptosis [92]. JNK-dependent apoptosis is inhibited by ERK-MAPK activation [93]. JNK can induce Bax phosphorylation, promote mitochondrial translocation, promote apoptosis, and inactivate Bcl-2 [94]. In BPH, there is an increased activity of the ERK cascade and a suppressive effect on the JNK and p38 pathways [67,95]. This is where the hormonal imbalance involved in BPH comes into play. Overexpression of EGF and IGF proteins has been observed in BPH [96].

It is assumed that excess DHT leads to the production of growth factors, particularly epidermal growth factor (EGF) and insulin growth factor (IGF) [97]. In prostate tissue, EFG and IGF receptor activation cause ERK/MAPK activation and consequently inhibit JNK-promoted apoptosis [98]. Transforming growth factor beta (TGF-β) is another growth factor that plays a beneficial role in BPH by inhibiting the proliferation process and promoting apoptosis in epithelial cells [99]. TGF-β activity can be influenced by DHT [67,97].

Based on the previews available to date, an animal model with BPH demonstrated the anti-BPH activity of PSs extracted from *Cucurbita pepo* by regulating the balance between proliferative and apoptotic processes. The study demonstrated that administration of PSs increased TGF-β expression and prevented ERK activation, thus promoting apoptosis through caspase 3 activation due to JNK and p38 phosphorylation [67]. According to another study, *Serenoa repens* extract enriched with 3% SIT possesses pro-apoptotic action in the prostate tissue of rats by inhibiting pAkt [78]. Akt, also known as protein kinase B, phosphorylates threonine and serine residues in target proteins. Akt activation increases the expression of anti-apoptotic stimuli (Bcl-2) through cAMP response element-binding protein (CREB) [100]. In addition to the decrease in the activity of the pro-apoptotic protein Bcl-2, PSs treatment led to an increase in the expression of Bax and procaspase-9. Conversely, another in vivo study reported no changes in the Bax/Bcl-2 ratio after PSs administration [67]. Furthermore, SIT has been reported to exert pro-apoptotic effects in prostate cancer by activating the sphingomyelin cycle [101]. It is well known that PSs inhibit the absorption of cholesterol by competing with it to enter the cell. This process activates Sph synthetase, an enzyme that favors the production of ceramides. Accumulation of ceramides leads to the activation of phosphatase A (PP2A), an enzyme that inhibits Akt (Figure 3).

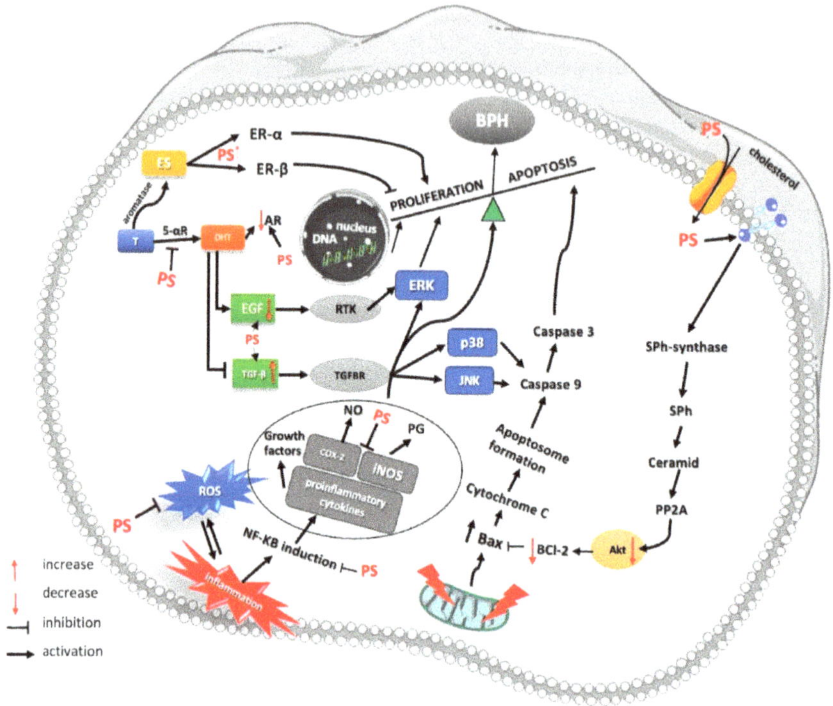

Figure 3. Schematic representation of mechanism involved in BPH. Anti-androgenic, anti-inflammatory, antioxidant, and pro-apoptotic mechanisms of PSs. PS: phytosterols, ES: estrogen, T: testosterone, DHT: dihydrotestosterone, Bcl-2: B-cell lymphoma 2, Bax: Bcl-2- associated X protein, ERK: extracellular signal-regulated protein kinase, JNK: Jun N-terminal Kinase, EGF: epidermal growth factor, TGF-β: transforming growth factor β, PP2A: phosphatase A, 5-αR: 5α-reductase, COX-2: cyclooxygenase-2, iNOS: inducible nitric oxide synthase, RTK: receptor tyrosine kinase, TGFBR: TGF-β receptor, ROS: reactive oxygen species; PG: prostaglandins; Akt: protein kinase B; SPh: sphingomyelin; Sph-synthase: sphingomyelin synthase.

3.3. Phytosterols in BPH—Clinical Evidence

PSs have been studied in a limited number of clinical trials in men with LUTS caused by BPH, but the results are inconclusive. In a randomized, double-blind, placebo-controlled multicenter trial in patients with BPH, patients received SIT (which contained a mixture of PSs) three times/day or a placebo for 6-months. It was shown that SIT was effective, as evidenced by significant improvements in urinary symptoms, urinary flow measures, and QL. There was no significant decline in prostatic volume in either the SIT or placebo groups [102]. In addition to the 6-month trial, an 18-month follow-up study was the only study that examined the long-term effects of PSs on patients with BPH. After 18 months, SIT treatment was continued to provide beneficial effects [103]. Another clinical study reported similar results when using SIT to treat BPH, showing slightly faster changes than those reported by Berges et al. [104]. The studies discussed are summarized in Table 2.

Several plant species were used in the studies as purified extracts, and the dosages ranged from 0.15 mg/day to 130 mg/day. None of the studies specified the use of PSs in a standardized extract form; for a review, see [102–108]. Using plant extracts containing different SIT dosages may pose problems when combining studies. In these trials, SIT concentrations should be known through the use of standardized extracts. Moreover, the only study that used 100% PSs did not show improvements in men with LUTS attributable to BPH [107]. Another randomized controlled trial demonstrated that *Serenoa repens* extract enriched with 3% PSs has superior efficacy compared to the simple *Serenoa repens* extract in relieving BPH symptoms, thus underlying the importance of PSs [109].

Table 2. Clinical studies with PSs effects in HBP.

Type of Study and Subjects	Treatment	Outcomes	Ref.
RCT patients with HBP (n = 200). Duration: 6 months	SIT (20 mg, which contains a mixture of PS), three times/day or placebo	Significant improvement in symptoms score and urinary flow parameters	[102]
RCT patients with HBP (n = 117) Duration: 18 months	SIT (20 mg, which contains a mixture of PS), three times/day or placebo	The effects on QOL of SIT are maintained over at least 18 months in men with symptomatic BPH	[103]
RCT patients with HBP (n = 177) Duration: 6 months	SIT (130 mg) and placebo	SIT is an effective option in the treatment of BPH.	[104]
Single-site study Randomization: noted but method not described (n = 62) Duration: 6 months	SIT (0,15 mg) and placebo	No improvements	[107]
Single-site study Randomization: unclear (n = 80) Duration: 1 month	SIT (65 mg) and placebo	SIT is an effective option in the treatment of BPH	[108]
RCT mild-to-moderate BPH symptoms (n = 99) Duration: 3 months	Saw palmetto oil (3% SIT), Saw palmetto oil (0.2% SIT) and placebo	Efficacy of SIT enriched saw palmetto oil is superior to conventional oil	[109]

The available data from these studies suggest that SIT improves urinary symptoms, flow measures, and QL. All studies were double-blinded and placebo-controlled. The long-term effects of phytosterols on symptomatology or QL have not been extensively studied in studies beyond 18 months. In addition, there is a lack of comparison with conventional therapies (alpha 1-adrenergic receptor antagonists, 5α-reductase inhibitors, or other natural compounds), which are the most widely used and effective drugs for the treatment of BPH. However, to confirm these findings, larger population studies with more robust methodologies are needed.

4. Regulatory Framework and Agreement with Label Declaration

Despite extensive research on DS, little attention in public health has focused on challenges in their regulation. There is no evidence of additional benefits with consumption above 3 g/day, and high consumption may have undesirable effects, according to the Scientific Committee on Food of the European Commission; therefore, it is prudent to avoid the consumption of plant sterols of more than 3 g/day [110]. Based on the data reviewed, PSs are generally well tolerated, with few side effects. Several clinical studies have reported adverse reactions to PSs intake, such as flatulence, diarrhea, constipation, nausea, and indigestion [102,111]. Another study demonstrated the safety of long-term consumption of plant sterols (one year) [112].

Serenoa repens extract is the main source of PSs used in BPH and is a relatively expensive extract; therefore, it is often falsified [113]. European Union (EU) legislation does not include specific regulations for DS-containing PSs. The United States Pharmacopoeia (USP) requires as an acceptance criterion that the *Serenoa repens* extract contains no less than 0.2%

total phytosterols and 0.1% SIT [114]. The same provisions are mentioned in the European Pharmacopoeia (Ph. EUR) [115].

The effective daily dose of PSs for the prevention or treatment of BPH has not been established by any regulatory framework. However, based on a few clinical studies, it is estimated that a dose between 0.060 and 0.140 g of SIT improves symptoms associated with BPH [102–105,116]. To date, only a few quantitative screening studies of food supplements containing PSs have been conducted. The amounts determined for each study are summarized in Table 3.

Table 3. Comparison of levels of phytosterols in dietary supplements between different studies.

Phytosterols Percentage of Sterol in Total Extract % (*w/w*)				
β-Sitosterol	**Stigmasterol**	**Campesterol**	**Total Sterols**	**References**
0.030–0.200	0.004–0.044	0.009–0.062	0.055–0.308	[117]
0.249–0.271	0.081–0.166	0.057–0.144	0.420–0.585	[118]
0.141–0.388	0.011–0.136	0.038–0.155	NR	[119]
0.010–3.255	0.001–1.975	0.003–2.228	0.015–7.511	[120]
0.001–0.276	0.008–0.104	0.022–0.117	NR	[121]
0.1	NR	NR	0.2	EU/USP monograph [114,115]
0.041	0.004	0.010	NR	SRM 3250 [39]
0.164	0.022	0.050	NR	SRM 3251 [39]

NR—not reported.

The estimated amount of PSs, specifically SIT, in DS-PSs-based treatment for BPH in our reviewed data was in accordance with the estimated amount needed to trigger beneficial effects. Based on the collected data (Table 3), it is evident that the amount of PSs in different DS varied significantly, possibly explained by the lack of a standardized extract of the active principle. In addition, the variability in the content may also be due to the analytical techniques used. The studies listed above used different methods of processing and quantifying samples, except for the study conducted by Sorenson et al. [119]. This was a collaborative study among ten laboratories that analyzed the PSs content of four DS using the same analytical method. Different manufacturers' PSs contain varying levels of PS but not those produced by the same company [120]. In a study by Cheon Kim et al., the DS content was compared with that of a standardized extract of *Serenoa repens* [118]. PSs levels were 1.4–2 times higher than the reference values. According to the authors of the aforementioned study, the products were adulterated, perhaps by adding less quality extract and sterol-rich exogenous vegetable lipid fractions [118]. Simultaneously, according to the USP and Ph. EUR regulations, not all supplements meet the acceptance criteria [118]. Moreover, in a study conducted by Penugonda et al., DS that exceeded the permissible daily dose of 3 g/day were identified, with a reported maximum PSs of 7.511 g [120]. Other studies have reported the lack or presence of undetectable amounts of plant sterols in some DS with declared PSs content [122].

5. Analysis of Phytosterols in Dietary Supplements

5.1. Sample Preparation

The first step in characterizing the supplement products is generally represented by the sample preparation. PSs are minor components that typically comprise less than 1% of the matrix (but up to 8% in foods with added PSs or DS). The objectives of this first step are to isolate the sterol fraction and convert all CPs into FPS for final analysis [26].

The extraction method should be selected according to the nature of the matrix, the physical properties of the sample (powder, solution, tincture, etc.), and the form (free or conjugated) in which the plant sterols are found. If we discuss DS, the matrix is a complex one containing both FPS and CPS, but also other organic compounds with similar structures that are extractable by saponification/solvent extraction, such as tocopherols, retinol, and β-carotenes, which can interfere with the quantification of PSs [123]. However, in practice, their levels are low compared to those of PSs, and their effect on the quantification of plant sterols remains statistically uncertain [124].

The DS processing methods reported in the literature are listed in Table 4. The first step involved extracting the lipid fraction from the matrix. Extraction may be carried out using liquid-liquid extraction (LLE) with organic solvents: hexane [125,126], heptan [123], chloroform [127,128], petroleum ether [22,129], or combinations of solvents with different polarities (chloroform-methanol, chlorophorm-methanol-water) [39,130]. This method has the advantages of using simple equipment and low cost of analysis. Solid-phase extraction (SPE) is another commonly used method and is a newer (eco-friendly) technique that provides a more convenient way to extract PSs with minimal analyte loss. Both LLE and SPE can result in recovery yields of 95–100% [131]. Recovery studies should be introduced into the extraction procedures of the analyte of interest to determine the effectiveness of the extraction. The available data also provide other modern extraction procedures suitable for PSs, such as supercritical fluid extraction, microwave-assisted extraction, ultrasound-assisted extraction, ionic liquid extraction, and enzyme-assisted extraction, according to the analyte matrix of interest [132].

Owing to the lack of commercial standards for CPS, an intermediate stage involves saponification at room temperature or directly by heating to obtain FPS. Saponification with an ethanol solution of KOH 2M at 80 °C for 30–60 min was noted to be the most commonly used [122]. Sometimes, the glycoside bond in the glycosylated form cannot be hydrolyzed under alkaline conditions, and acid hydrolysis is required [133,134], as shown in Figure 4. The hydrolysis process is followed by the extraction of unsaponifiable lipids by LLE or SPE. Sample processing by alkaline saponification and direct acid hydrolysis without prior extraction of the lipid fraction has also been shown to be an effective method in quantifying the FPS and CPS of *Saccharum officinarum*, using reverse-phase high-performance chromatography (RP-HPLC) as a method of determination [135].

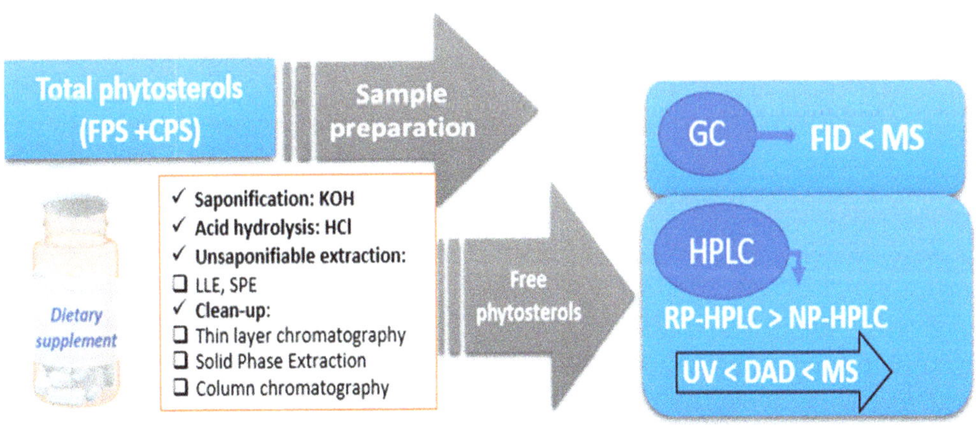

Figure 4. Main steps in the determination of total phytosterols. IS (internal standard) TMS (Trimethylsilyl ether); HPLC (high liquid chromatography); GC (gas chromatography); NP (normal phase); RP (reverse phase).

In the final step, the removal of potentially interfering compounds (aliphatic alcohols), isolation, and concentration of the extracts involves methods such as thin-layer chromatography (TLC) or column chromatography (CC). As these analytical methods are time-consuming, they have been replaced by SPE or solid-phase microextraction [131,136]. The main sample preparation sequence is shown in Figure 4.

5.2. Determination Methods

A variety of chromatographic methods have been used to characterize and quantify plant sterols isolated from different samples, including gas chromatography (GC) [22,118–120,127,128,137,138], thin-layer chromatography (TLC) [23,131,139], and high-performance liquid chromatography (HPLC) [39,122,140–143]. In addition to classical methods, Fourier transform infrared spectroscopy (FT-NIST), based on previously known reference values obtained by conventional methods, has been introduced for the rapid analysis of total and individual sterols; for a review, see [144–146]. Furthermore, Gomez SM et al. reported a physical approach for the quantitative analysis of free sterols and their mixtures in vegetable oils using X-ray powder diffraction and the Rietveld method [147]. In the following sections, the main analytical methods used for monitoring PSs in DS are discussed.

- Gas Chromatography

GC is the standard method for quantifying PSs; for a review, see [23,131,146]. As shown in Table 4, most of the identified studies used GC to analyze DS with PSs content. Instead, for structural identification, GC coupled with mass spectrophotometry (MS) with chemical or electron ionization is an appropriate method for the simultaneous acquisition of both the retention times and molecular masses of the components in the mixture [148]. Furthermore, GC-MS eliminates the problem of co-elution of compounds of interest, which is frequently encountered when determining PSs [22,149].

However, GC presents a laborious sample processing method, including the derivatization of analytes with the formation of trimethylsilyl ethers (TMS) or acetylated derivatives to favor the volatilization of plant sterols [23,146]. The most commonly used derivatizing agents are N-methyl-N-(trimethylsilyl) fluoroacetamide (MTSFA) in anhydrous pyridine and bis-(trimethylsilyl) trifluoroacetamide (BSTFA) with 1% trimethylchlorosilane (TMCS) [23]. The derivatization reaction is difficult to validate because other compounds in the mixture may interact with the derivatization agent. In addition, some degradation products of the derivatization agent may interfere with the chromatographic signal [122]. Furthermore, GC has the disadvantage of requiring high temperatures for analysis, which are not thermally suitable for unstable products, such as plant sterols [122,150,151]. To overcome the effects of fluctuations in instrument operation conditions and other experimental variables, the GC chromatographic peak of a plant sterol is compared to an external or internal standard. Moreover, the correction of the losses of the analyte of interest in quantitative determination during sample processing can be performed using the internal standard [131]. The surrogate internal standards mainly used for quantification of PSs by GC-FID are betuline, 5α-colestanol, and 5β-colestan-3α-ol (epicoprostanol) [23,131].

Due to the structural similarities between PSs and cholesterol, they can be extracted and chromatographically analyzed under similar conditions. Therefore, the Association of Chemical Analysis (AOAC) Official Method 994.10, "Cholesterol in Foods," has been modified and validated for this purpose. The samples were saponified at high temperatures with ethanolic KOH, and the unsaponifiable fractions (SIT, stigmasterol, campesterol) were extracted with toluene. PSs were derivatized to TMS and quantified using GC-FID [121]. Later, an inter-laboratory collaborative study evaluated and recommended the method described above for the determination of PSs from DS at concentrations between 1 and 5 mg/100 g [119].

Currently, the AOAC does not offer an official method for the determination of PSs in different foods, and the previously described determinations have not been validated for the quantification of PSs in fortified/enriched foods and DS with content greater than 1% PSs and do not provide for the identification of plant stanols [119,121]. The methods developed and validated by Laakso et al. and Clement et al. were considered as possible replacement methods, being validated in a single laboratory for a wide range of PSs in a broad range of linearity. Unfortunately, neither method provided adequate GC separation of phytosterols and phytostanols [152,153].

In spite of this, a validated method is available, which uses the traditional sample processing method for the identification of 16 PSs, including the most important ones (SIT, stigmasterol, campesterol) from foods and DS enriched with PSs. This method was adapted for DS as follows: for DS with content greater than 8 g/100 g, it is necessary to use samples smaller than 0.5 g and to increase the amount of solvent for the extraction of the unsaponified fraction (10 mg/mL concentration for total PSs). The application range was between 0.001 g PS/100 g (quantification limit)—55.2 g PS/100 g [22]. The GC methods described in the literature for the determination of sterols from different food matrices or plant products have been published and reviewed by Abdi et al. [131], Lagarda et al. [23], and Garcia-Latas et al. [146].

- High-performance Liquid Chromatography

As an alternative to GC, LC has also been used to determine PSs from various matrices, including DS (Table 4). PSs detection was performed using simple UV [122,142,154] or diode array detection (DAD) [141,155–157], corona-charged aerosol detector (CAD) [122,158,158], refractive index (RI) detection [26,146], evaporative light scattering detection (ELSD) [159,160], fluorescence (FL) [161], nuclear magnetic resonance (NMR) [26,146,162], and mass spectrometry (MS) [23,39,141,163]. Reversed-phase HPLC (RP-HPLC) is commonly used for PSs analysis and separation as opposed to normal-phase HPLC (NP-HPLC). For the reversed stationary phase, silica gel supports containing octadecyl (C18)-linked alkyl groups are usually used, whereas the mobile phase is composed of acetonitrile, methanol, or a mixture of water and organic solvents [23,131].

Compared to GC, HPLC methods do not require derivatization, which is an expensive and time-consuming process and can sometimes interfere with the detection method, thus causing the loss of the analyte of interest. However, derivatization is sometimes necessary to optimize the detector response by improving the separation and ionization efficiency in LC-MS [161].

A further advantage of LC is that it operates at lower temperatures, often at room temperature, making it an ideal method for examining thermally labile compounds, such as PSs; for a review, see [23,122,163], and could be performed by using a non-destructive detector, compared to the FID detector in the GC.

On the other hand, HPLC analysis is associated with many difficulties. If the matrix used is simple and homogeneous (e.g., seed oil), direct HPLC analysis can be applied for the determination of PSs without the laborious step of sample processing. Otherwise, if the analyte of interest is in a complex matrix, preliminary sample purification methods (CC, TLC) are required [131]. Due to the structural similarities, PSs are difficult to separate, requiring increased running time to avoid co-elution. Longer retention times are often associated with the broadening of the chromatographic peaks [122]. Moreover, HPLC-UV or DAD analysis of PSs is associated with sensitivity problems due to the lack of chromophore groups and the broadening of peaks associated with their high lipophilicity at high retention times in RP-HPLC. PSs absorb at wavelengths between 200 and 210 nm. These low wavelengths are not selective; therefore, interference can be observed in the chromatograms by revealing interfering compounds, which may be present in the samples after extraction [145,157].

Recently, a new analytical method, UHPLC with tandem DAD (UV)/charged aerosol detector (CAD), allows rapid determination (<8.5 min) and efficient chromatographic separation of PSs from DS. It has been demonstrated that CAD sensitivity is three times greater than that of UV at wavelengths below 210 nm [122].

Table 4. Determination of total phytosterols in dietary supplements by chromatographic methods.

Method	Detector	Sample Preparation/Extraction	Analytical Methods	Target Compounds	Ref.
GC	FID	• Addition internal standard solution • Saponification: KOH (2 h/100 °C) • Derivatization: BSTFA, and pyridine	Column: (25m × 0.25mm, 0,33 µm, Dimethylpolysiloxane stationary phase) Temperatures (°C): Detector: 325; Injector: 325 Gas (mL/min): Helium (0.5) Injection type: Split ratio, 1:40 Injection volume (µL): 1	campesterol stigmasterol SIT total sterols	[118]
GC	FID/MS	• Addition internal standard solution (epicoprostanol) • Acid hydrolysis HCl (4M/1 h/100 °C) • Lipid extraction: EE: diethyl ether: petroleum ether (50:50/*v:v*) • Saponification: ethanolic KOH (2 M/80 °C/60 min) • Unsaponifiable extraction (LLE): EE (3 times) • Derivatization: BSTFA, and pyridine	Column: (30 m × 0.32 mm, 0.25 µm, (poly (94%methyl/5% phenyl) silicone)) Temperatures (°C): column 250, injector 290, detector 290 Gas (mL/min): H2 (1) Injection type Split ratio, 1:25 Injection volume (µL): 1	campesterol stigmasterol SIT	[22]
GC	FID	• Alkaline and acid/alkaline protocols • Addition internal standard solution (epicoprostanol) • Acid hydrolysis: HCl (3N) • Saponification: NaOH 2.3 N in MeOH + HCl 3N + NaCl • Derivatization: BSTFA, and pyridine	Column: (30 m × 0.32 mm, 0.25 µm) (5%-phenyl)-methylpolysiloxane stationary phase)/(30 m × 0.32 mm, 0.25 µm, (5%-phenyl) (1%-vinyl)-methylpolysiloxane stationary phase) Temperatures (°C): column 250, Injector 290, detector 290 Gas (mL/min): H2 (1) Injection type Split ratio, 1:25 Injection volume (µL): 1	17 phytosterols	[138]
GC	FID/MS	• Addition internal standard solution (epicholesterol) • Saponification: aqueous KOH (1.3 N/85–89 °C/20 min) • Unsaponifiable extraction: cyclohexane • Derivatization: (BSTFA/TMCS)	Column: (15 m × 0.32-mm,0.25 µm/5% diphenyl-95%-Dimethylpolysiloxane) Temperatures (°C): column 270, injector 280, detector 300. Gas (ml/min): helium (0.58) Injector type: split ratio, 17:1 Injection volume (µL): 0.5	SIT β-sitosteryl glucoside (BSSG)	[139]

Table 4. *Cont.*

Method	Detector	Sample Preparation/Extraction	Analytical Methods	Target Compounds	Ref.
GC	FID	• Saponification: ethanolic KOH (1 M/hot plate/80–90 min) • Unsaponifiable extraction: toluene • Derivatization: TMCS • Addition internal standard solution (5 alfa-cholestane)	Column: (25 m × 0.32 mm, 17 μm/5% phenyl-methylsilicone or methyl silicone gum stationary phase) Temperatures (°C): Column 190→255, injector: 250, detector: 300 Gas (mL/min): helium 2/15/3; makeup helium 20/hydrogen 35/air 380 Injection volume (μL): 1	Campesterol Stigmasterol SIT	[119]
GC	FID/MS	• Addition internal standard solution (cholestanol), chloroform • Saponification: methanol KOH (0.3M/80 °C/60 min) • Unsaponifiable extraction: hexane (twice) • Derivatization: MTSA, TMCS, and pyridine	Column: (60 m × 0.25 mm, 0.25 um/5% phenyl methyl siloxane stationary phase) Temperatures (°C): column 80→325, detector 230. Gas (mL/min): helium (1) Injection volume (μL): 1	campesterol stigmasterol SIT total sterols	[120]
LC	APCI–MS	• Solvent extraction: hexane • Saponification with base (ethanolic KOH)/acid hydrolysis • Solvent extraction of unsaponifiable material: toluene	1. Column: phenyl (150 mm × 3.9 mm, 3,5 μm) Mobile phase: 58% acetonitrile, 42% water Flow rate (mL/min): 1.1 (isocratic) 2. Column: ACE C18 (150 mm × 3.0 mm, 3 μm) Mobile phase: 90% methanol, 10% water Flow rate (mL/min): 0.80 (isocratic) Injection volume (μL): 5 for qualitative, 10 for quantitative measurement	campesterol, cycloartenol, lupenone, lupeol, SIT, and stigmasterol (Standard Reference Materials containing saw palmetto)	[39]
HPLC	ESI-MS	• Saponification: ethanolic KOH (70 °C/30 min) • Unsaponifiable extraction: hexane • Derivatization: dansyl chloride (40 °C/30 min)	Column: C18 (250 × 3.0 mm, 5 μm) Mobile phase: 95% methanol: 5% water Flow rate (mL/min): 0.5 (isocratic)	campesterol stigmasterol β-sitosterol	[141]
HPLC	UV	• Ultrasound-Assisted Emulsification • Microextraction (USAEME): • Solubilization: methanol • Addition: calcium chloride 2mol/L • Fat extraction: hexane (heating for 10 min at 50 °C/ultrasonic time 25 min/centrifugation 2600 rpm, 15 min)	Column: RP-18 (125-4 mm, 5 μm) Mobile phase: hexane, propan-1-ol (99.5:0.5; *v*/*v*) Flow rate (mL/min): 0.8–1 (isocratic) Injection volume (μL): 50 UV detection: 212 nm	β-sitosterol	[142]

Table 4. *Cont.*

Method	Detector	Sample Preparation/Extraction	Analytical Methods	Target Compounds	Ref.
UHPLC	UV/CAD	• Addition internal standard solution (cholesterol) • Saponification: ethanolic KOH (80 °C/30 min) • Acid hydrolysis: HCl • Solvent extraction: hexane (3 min)	Column: (100 × 2.1 mm, 1,7 µm) Phenyl-hexyl Mobile phase: acetonitrile: water Flow rate (mL/min): 0.9 (gradient) Temperature: 60 °C Inj. vol. (µL): 2 Run time (min): 8.5	ergosterol, brassicasterol, campesterol, campestanol, fucosterol, stigmasterol, stigmastanol, SIT esterified form	[122]

GC: gas chromatography; FID: flame ionization detection; BSTFA: bis-(trimethylsilyl) trifluoroacetamide; LLE: Liquid-liquid extraction; MS: mass spectrometry; TMCS: trimethylchlorosilane; MTSFA: N- methyl-N-(trimethylsilyl) fluoroacetamide; LC-liquid chromatography; APCI: atmospheric pressure chemical ionization; ESI: ionization by electrospray; UV: ultraviolet; HPLC: high liquid chromatography; UHPLC: ultra-high liquid chromatography; CAD: Charged aerosol detector.

However, LC-MS can address many of the challenges encountered in previous analytical strategies due to the selectivity and specificity of the detection method [23,163]. The most common types of MS detectors for PSs analysis based on the ionization source are electrospray ionization (ESI), atmospheric pressure chemical ionization (APCI), atmospheric pressure photoionization (APPI), matrix-assisted laser desorption ionization (MALDI), and ambient MS, such as direct analysis in real time (DART) [163]. Because PSs are extremely lipophilic with few polar functional groups, ESI, which is one of the most widespread and powerful ionization techniques, is unsuitable for their determination [23,141,164,165]. Thus, other ionization techniques, such as APCI or APPI [39,141,166], were applied. On the other hand, quantitative MS-based analysis from complex matrices is associated with an unpredictable matrix effect and increased cost of the internal standard [167].

A recent paper reported a fast, simple, and low-cost dansylation derivatization method that solves previously encountered problems. The optimal derivatization reaction conditions consisted of dichloromethane as the solvent and 4-dimethylaminopyridine as the catalyst at 40 °C for 30 min. The derivatization process by dansylation allows the improvement of UV detection (254 nm), the limitation of the detection of non-target compounds, and the separation by RP-HPLC, with the exception of stigmasterol and campesterol. Moreover, this method solves the problem of ESI inefficiency, which leads to increased reproducibility and linearity [141].

6. Conclusions

This review summarizes the potential effects of PSs intake on BPH by presenting the thus far published data obtained from in vitro studies, animal studies, and clinical trials. BPH prevention and treatment require a detailed understanding of the molecular mechanisms underlying PSs-induced activities. The use of PSs for BPH is primarily explained by their anti-androgenic activity through 5-αR inhibition but can also act via apoptotic pathways linked to the endocrine system. Recently, the anti-inflammatory and antioxidant actions of PSs have emphasized their potential for use as individual active principles in BPH. Considering that the quantity and quality of PSs in DS are intrinsically related to their efficacy, the data on monitoring PSs content in DS has been reviewed. Substantial variability in botanical supplement composition and concentrations has been noted. In the future, the use of standardized, rapid, and economic analysis techniques may allow the implementation of a system to certify the authenticity of PSs content in DS used in therapy. Overgeneralization of pharmacological effects of all plant-mixture without any specific emphasis on PSs compounds and the shortage of any valid regulatory framework is also considered a significant challenge.

To conclude, PSs should be considered with other medical therapies for patients with symptomatic BPH, but lack of standardization and pharmacological effect limits the potential clinical usefulness. Thus, further studies are needed to ensure a more in-depth understanding of the molecular mechanisms underlying PSs-induced activities and to design upcoming strategies to overcome the currently identified regulation and analytical-related gaps.

Author Contributions: Writing—original draft preparation, M.-G.B., writing-review and editing C.-M.J., B.E.Ő., G.J., C.E.V. and S.I.; supervision, A.T.-V. All authors have read and agreed to the published version of the manuscript.

Funding: This research received no external funding.

Institutional Review Board Statement: Not applicable.

Informed Consent Statement: Not applicable.

Data Availability Statement: Not applicable.

Conflicts of Interest: The authors declare no conflict of interest.

References

1. Wei, J.T.; Calhoun, E.; Jacobsen, S.J. Urologic diseases in America project: Benign prostatic hyperplasia. *J. Urol.* **2005**, *173*, 1256–1261. [CrossRef] [PubMed]
2. Parsons, J.K.; Bergstrom, J.; Silberstein, J.; Barrett-Connor, E. Prevalence and characteristics of lower urinary tract symptoms in men aged > or =80 years. *Urology* **2008**, *72*, 318–321. [CrossRef] [PubMed]
3. Kim, T.H.; Lim, H.J.; Kim, M.S.; Lee, M.S. Dietary supplements for benign prostatic hyperplasia: An overview of systematic reviews. *Maturitas* **2012**, *73*, 180–185. [CrossRef]
4. Eleazu, C.; Eleazu, K.; Kalu, W. Management of Benign Prostatic Hyperplasia: Could Dietary Polyphenols Be an Alternative to Existing Therapies? *Front. Pharmacol.* **2017**, *8*, 234. [CrossRef]
5. Fogaing, C.; Alsulihem, A.; Campeau, L.; Corcos, J. Is Early Surgical Treatment for Benign Prostatic Hyperplasia Preferable to Prolonged Medical Therapy: Pros and Cons. *Medicina* **2021**, *57*, 368. [CrossRef] [PubMed]
6. Peng, Z.F.; Zhou, J.; Song, P.; Yang, L.C.; Yang, B.; Ren, Z.J.; Wang, L.C.; Wei, Q.; Dong, Q. Retrospective analysis of the changes in the surgical treatment of benign prostatic hyperplasia during an 11-year period: A single-center experience. *Asian J. Androl.* **2021**, *23*, 294–299. [CrossRef]
7. Tarter, T.H.; Vaughan, E.D., Jr. Inhibitors of 5alpha-reductase in the treatment of benign prostatic hyperplasia. *Curr. Pharm. Des.* **2006**, *12*, 775–783. [CrossRef]
8. Kim, E.H.; Brockman, J.A.; Andriole, G.L. The use of 5-alpha reductase inhibitors in the treatment of benign prostatic hyperplasia. *Asian J. Urol.* **2018**, *5*, 28–32. [CrossRef]
9. Azzouni, F.; Godoy, A.; Li, Y.; Mohler, J. The 5 alpha-reductase isozyme family: A review of basic biology and their role in human diseases. *Adv. Urol.* **2012**, *2012*, 530121. [CrossRef]
10. Kim, J.H.; Baek, M.J.; Sun, H.Y.; Lee, B.; Li, S.; Khandwala, Y.; Del Giudice, F.; Chung, B.I. Efficacy and safety of 5 alpha-reductase inhibitor monotherapy in patients with benign prostatic hyperplasia: A meta-analysis. *PLoS ONE* **2018**, *13*, e0203479. [CrossRef]
11. Gormley, G.J.; Stoner, E.; Bruskewitz, R.C.; Imperato-McGinley, J.; Walsh, P.C.; McConnell, J.D.; Andriole, G.L.; Geller, J.; Bracken, B.R.; Tenover, J.S. The effect of finasteride in men with benign prostatic hyperplasia. The Finasteride Study Group. *N. Engl. J. Med.* **1992**, *327*, 1185–1191. [CrossRef] [PubMed]
12. Kim, E.H.; Larson, J.A.; Andriole, G.L. Management of Benign Prostatic Hyperplasia. *Annu. Rev. Med.* **2016**, *67*, 137–151. [CrossRef] [PubMed]
13. Kaplan, S.A.; Lee, J.Y.; Meehan, A.G.; Kusek, J.W. Time Course of Incident Adverse Experiences Associated with Doxazosin, Finasteride and Combination Therapy in Men with Benign Prostatic Hyperplasia: The MTOPS Trial. *J. Urol.* **2016**, *195*, 1825–1829. [CrossRef]
14. Roehrborn, C.G.; Boyle, P.; Nickel, J.C.; Hoefner, K.; Andriole, G. Efficacy and safety of a dual inhibitor of 5-alpha-reductase types 1 and 2 (dutasteride) in men with benign prostatic hyperplasia. *Urology* **2002**, *60*, 434–441. [CrossRef]
15. Mysore, V. Finasteride and sexual side effects. *Indian Dermatol. Online J.* **2012**, *3*, 62–65. [CrossRef]
16. The National Institutes of Health (NIH). Dietary Supplement Label Database. Available online: https://dsld.od.nih.gov/search/prostate/bWFya2V0X3N0YXR1cz1hbGwvZW50cnlfZGF0ZT0yMDExLDwMjIvc29ydD1tYXRjaC9wYWdlX3NpemU9MjAvcGFnZV9pbmRleD0xLw (accessed on 7 February 2022).
17. Leisegang, K.; Jimenez, M.; Durairajanayagam, D.; Finelli, R.; Majzoub, A.; Henkel, R.; Agarwal, A. A Systematic Review of Herbal Medicine in the Clinical Treatment of Benign Prostatic Hyperplasia. *Phytomedicine* **2022**, *2*, 100153. [CrossRef]
18. Wilt, T.J.; Ishani, A.; Rutks, I.; MacDonald, R. Phytotherapy for benign prostatic hyperplasia. *Public Health Nutr.* **2000**, *3*, 459–472. [CrossRef]
19. Csikós, E.; Horváth, A.; Ács, K.; Papp, N.; Balázs, V.L.; Dolenc, M.S.; Kenda, M.; Kočevar Glavač, N.; Nagy, M.; Protti, M.; et al. Treatment of Benign Prostatic Hyperplasia by Natural Drugs. *Molecules* **2021**, *26*, 7141. [CrossRef]
20. Hosbas Coskun, S.; Wise, S.A.; Kuszak, A.J. The Importance of Reference Materials and Method Validation for Advancing Research on the Health Effects of Dietary Supplements and Other Natural Products. *Front. Nutr.* **2021**, *8*, 786261. [CrossRef]
21. Bot, A. Phytosterols. In *Encyclopedia of Food Chemistry*; Melton, L., Shahidi, F., Varelis, P., Eds.; Academic Press: Oxford, UK, 2019; pp. 225–228. [CrossRef]
22. Srigley, C.T.; Haile, E.A. Quantification of Plant Sterols/Stanols in Foods and Dietary Supplements Containing Added Phytosterols. *J. Food Compost. Anal.* **2015**, *40*, 163–176. [CrossRef]
23. Lagarda, M.J.; García-Llatas, G.; Farré, R. Analysis of phytosterols in foods. *J. Pharm. Biomed. Anal.* **2006**, *41*, 1486–1496. [CrossRef] [PubMed]
24. García-Llatas, G.; Rodríguez-Estrada, M.T. Current and New Insights on Phytosterol Oxides in Plant Sterol-Enriched. *Food. Chem. Phys. Lipids* **2011**, *164*, 607–624. [CrossRef] [PubMed]
25. Moreau, R.A.; Whitaker, B.D.; Hicks, K.B. Phytosterols, phytostanols, and their conjugates in foods: Structural diversity, quantitative analysis, and health-promoting uses. *Prog. Lipid. Res.* **2022**, *41*, 457–500. [CrossRef] [PubMed]
26. Moreau, R.A.; Nyström, L.; Whitaker, B.D.; Winkler-Moser, J.K.; Baer, D.J.; Gebauer, S.K.; Hicks, K.B. Phytosterols and Their Derivatives: Structural Diversity, Distribution, Metabolism, Analysis, and Health-Promoting Uses. *Prog. Lipid Res.* **2018**, *70*, 35–61. [CrossRef] [PubMed]
27. Daels, E.; Foubert, I.; Goderis, B. The effect of adding a commercial phytosterol ester mixture on the phase behavior of palm oil. *Food Res. Int.* **2017**, *100 Pt 1*, 841–849. [CrossRef]

28. Davidson, M.H.; Maki, K.C.; Umporowicz, D.M.; Ingram, K.A.; Dicklin, M.R.; Schaefer, E.; Lane, R.W.; McNamara, J.R.; Ribaya-Mercado, J.D.; Perrone, G.; et al. Safety and tolerability of esterified phytosterols administered in reduced-fat spread and salad dressing to healthy adult men and women. *Am. Coll. Nutr.* **2001**, *20*, 307–319. [CrossRef]
29. Mussner, M.J.; Parhofer, K.G.; Von Bergmann, K.; Schwandt, P.; Broedl, U.; Otto, C. Effects of phytosterol ester-enriched margarine on plasma lipoproteins in mild to moderate hypercholesterolemia are related to basal cholesterol and fat intake. *Metabolism* **2002**, *51*, 189–194. [CrossRef]
30. Kwak, H.S.; Ahn, H.J.; Ahn, J. Development of Phytosterol Ester-Added Cheddar Cheese for Lowering Blood Cholesterol. *Asian-Australas J. Anim. Sci.* **2005**, *18*, 267–276. [CrossRef]
31. Piironen, V.; Lindsay, D.G.; Miettinen, T.A.; Toivo, J.; Lampi, A.-M. Plant sterols: Biosynthesis, biological function and their importance to human nutrition. *J. Sci. Food Agric.* **2000**, *80*, 939–966. [CrossRef]
32. Normén, L.; Bryngelsson, S.; Johnsson, M.; Evheden, P.; Ellegård, L.; Brants, H.; Andersson, H.; Dutta, P. The Phytosterol Content of Some Cereal Foods Commonly Consumed in Sweden and in the Netherlands. *J. Food Compos. Anal.* **2002**, *15*, 693–704. [CrossRef]
33. Othman, R.A.; Myrie, S.B.; Jones, P.J. Non-cholesterol sterols and cholesterol metabolism in sitosterolemia. *Atherosclerosis* **2013**, *231*, 291–299. [CrossRef]
34. Piironen, V.; Lampi, A.M. *Phytosterols as Functional Food Components and Nutraceuticals*; Dutta, P.C., Ed.; Marcel Dekker, Inc.: New York, NY, USA, 2004; pp. 1–32.
35. Martins, C.M.; Fonseca, F.A.; Ballus, C.A.; Figueiredo-Neto, A.M.; Meinhart, A.D.; de Godoy, H.T.; Izar, M.C. Common sources and composition of phytosterols and their estimated intake by the population in the city of São Paulo, Brazil. *Nutrition* **2013**, *29*, 865–871. [CrossRef]
36. Yang, R.; Xue, L.; Zhang, L.; Wang, X.; Qi, X.; Jiang, J.; Yu, L.; Wang, X.; Zhang, W.; Zhang, Q.; et al. Phytosterol Contents of Edible Oils and Their Contributions to Estimated Phytosterol Intake in the Chinese Diet. *Foods* **2019**, *8*, 334. [CrossRef] [PubMed]
37. Nair, P.P.; Turjman, N.; Kessie, G.; Calkins, B.; Goodman, G.T.; Davidovitz, H.; Nimmagadda, G. Diet, nutrition intake, and metabolism in populations at high and low risk for colon cancer. Dietary cholesterol, beta-sitosterol, and stigmasterol. *Am. J. Clin. Nutr.* **1984**, *40* (Suppl. S4), 927–930. [CrossRef]
38. Geavlete, P.; Multescu, R.; Geavlete, B. *Serenoa repens* extract in the treatment of benign prostatic hyperplasia. *Ther. Adv. Urol.* **2011**, *3*, 193–198. [CrossRef] [PubMed]
39. Bedner, M.; Schantz, M.M.; Sander, L.C.; Sharpless, K.E. Development of liquid chromatographic methods for the determination of phytosterols in Standard Reference Materials containing saw palmetto. *J. Chromatogr. A* **2008**, *1192*, 74–80. [CrossRef]
40. Schauer, I.G.; Rowley, D.R. The functional role of reactive stroma in benign prostatic hyperplasia. *Differentiation* **2011**, *82*, 200–210. [CrossRef] [PubMed]
41. Vickman, R.E.; Franco, O.E.; Moline, D.C.; Vander Griend, D.J.; Thumbikat, P.; Hayward, S.W. The role of the androgen receptor in prostate development and benign prostatic hyperplasia: A review. *Asian J. Urol.* **2020**, *7*, 191–202. [CrossRef] [PubMed]
42. Pizzorno, J.E.; Murray, M.T.; Joiner-Bey, H. 12—Benign Prostatic Hyperplasia. In *The Clinician's Handbook of Natural Medicine*, 3rd ed.; Pizzorno, J.E., Murray, M.T., Joiner-Bey, H., Eds.; Churchill Livingstone: Edinburgh, UK, 2016; pp. 137–146. [CrossRef]
43. Gao, W.; Bohl, C.E.; Dalton, J.T. Chemistry and structural biology of androgen receptor. *Chem. Rev.* **2005**, *105*, 3352–3370. [CrossRef]
44. Kalu, W.O.; Okafor, P.N.; Ijeh, I.I.; Eleazu, C. Effect of kolaviron, a biflavanoid complex from *Garcinia kola* on some biochemical parameters in experimentally induced benign prostatic hyperplasic rats. *Biomed. Pharmacother.* **2016**, *83*, 1436–1443. [CrossRef]
45. Ajayi, A.; Abraham, K. Understanding the Role of Estrogen in the Development of Benign Prostatic Hyperplasia. *Afr. J. Urol.* **2018**, *24*, 93–97. [CrossRef]
46. Miwa, Y.; Kaneda, T.; Yokoyama, O. Association Between Lower Urinary Tract Symptoms and Serum Levels of Sex Hormones in Men. *Urology* **2008**, *72*, 552–555. [CrossRef]
47. Ellem, S.J.; Risbridger, G.P. The dual, opposing roles of estrogen in the prostate. *Ann. N. Y. Acad. Sci.* **2009**, *1155*, 174–186. [CrossRef] [PubMed]
48. McPherson, S.J.; Hussain, S.; Balanathan, P.; Hedwards, S.L.; Niranjan, B.; Grant, M.; Chandrasiri, U.P.; Toivanen, R.; Wang, Y.; Taylor, R.A.; et al. Estrogen receptor-beta activated apoptosis in benign hyperplasia and cancer of the prostate is androgen independent and TNFalpha mediated. *Proc. Natl. Acad. Sci. USA* **2010**, *107*, 3123–3128. [CrossRef]
49. Robert, G.; Descazeaud, A.; Nicolaïew, N.; Terry, S.; Sirab, N.; Vacherot, F.; Maillé, P.; Allory, Y.; de la Taille, A. Inflammation in benign prostatic hyperplasia: A 282 patients' immunohistochemical analysis. *Prostate* **2009**, *69*, 1774–1780. [CrossRef]
50. Chughtai, B.; Lee, R.; Te, A.; Kaplan, S. Role of inflammation in benign prostatic hyperplasia. *Rev. Urol.* **2011**, *13*, 147–150.
51. Latil, A.; Pétrissans, M.T.; Rouquet, J.; Robert, G.; de la Taille, A. Effects of hexanic extract of *Serenoa repens* (Permixon® 160 mg) on inflammation biomarkers in the treatment of lower urinary tract symptoms related to benign prostatic hyperplasia. *Prostate* **2015**, *75*, 1857–1867. [CrossRef]
52. Nickel, J.C.; Roehrborn, C.G.; O'Leary, M.P.; Bostwick, D.G.; Somerville, M.C.; Rittmaster, R.S. The relationship between prostate inflammation and lower urinary tract symptoms: Examination of baseline data from the REDUCE trial. *Eur. Urol.* **2008**, *54*, 1379–1384. [CrossRef]
53. Kirschenbaum, A.; Klausner, A.P.; Lee, R.; Unger, P.; Yao, S.; Liu, X.-H.; Levine, A.C. Expression of Cyclooxygenase-1 and Cyclooxygenase-2 in the Human Prostate. *Urology* **2000**, *56*, 671–676. [CrossRef] [PubMed]

54. Hudson, C.N.; He, K.; Pascal, L.E.; Liu, T.; Myklebust, L.K.; Dhir, R.; Srivastava, P.; Yoshimura, N.; Wang, Z.; Ricke, W.A.; et al. Increased COX-1 expression in benign prostate epithelial cells is triggered by mitochondrial dysfunction. *Am. J. Clin. Exp. Urol.* **2022**, *10*, 234–245. [PubMed]
55. Bonkhoff, H. Estrogen receptor signaling in prostate cancer: Implications for carcinogenesis and tumor progression. *Prostate* **2018**, *78*, 2–10. [CrossRef]
56. Di Silverio, F.; Gentile, V.; De Matteis, A.; Mariotti, G.; Giuseppe, V.; Antonio Luigi, P.; Sciarra, A. Distribution of Inflammation, Pre-Malignant Lesions, Incidental Carcinoma in Histologically Confirmed Benign Prostatic Hyperplasia: A Retrospective Analysis. *Eur. Urol.* **2003**, *43*, 164–175. [CrossRef] [PubMed]
57. Minciullo, P.L.; Inferrera, A.; Navarra, M.; Calapai, G.; Magno, C.; Gangemi, S. Oxidative stress in benign prostatic hyperplasia: A systematic review. *Urol. Int.* **2015**, *94*, 249–254. [CrossRef]
58. Winner, K.; Polycarp, O.; Ifeoma, I.; Chinedum, E. Effect of Fractions of Kolaviron on Some Indices of Benign Prostatic Hyperplasia in Rats: Identification of the Constituents of the Bioactive Fraction Using GC-MS. *RSC Adv.* **2016**, *6*, 94352–94360. [CrossRef]
59. Cabeza, M.; Bratoeff, E.; Heuze, I.; Ramírez, E.; Sánchez, M.; Flores, E. Effect of beta-sitosterol as inhibitor of 5 alpha-reductase in hamster prostate. *Proc. West. Pharmacol. Soc.* **2003**, *46*, 153–155.
60. Kamei, H.; Noguchi, K.; Matsuda, H.; Murata, K. Screening of *Euphorbiaceae* Plant Extracts for Anti-5α-reductase. *Biol. Pharm. Bull.* **2018**, *41*, 1307–1310. [CrossRef]
61. Lesuisse, D.; Berjonneau, J.; Ciot, C.; Devaux, P.; Doucet, B.; Gourvest, J.F.; Khemis, B.; Lang, C.; Legrand, R.; Lowinski, M.; et al. Determination of oenothein B as the active 5-alpha-reductase-inhibiting principle of the folk medicine *Epilobium parviflorum*. *J. Nat. Prod.* **1996**, *59*, 490–492. [CrossRef] [PubMed]
62. Hartmann, R.W.; Mark, M.; Soldati, F. Inhibition of 5 α-reductase and aromatase by PHL-00801 (Prostatonin®), a combination of PY102 (*Pygeum africanum*) and UR102 (*Urtica dioica*) extracts. *Phytomedicine* **1996**, *3*, 121–128. [CrossRef] [PubMed]
63. Heim, S.; Seibt, S.; Stier, H.; Moré, M.I. Uromedic® Pumpkin Seed Derived Δ7-Sterols, Extract and Oil Inhibit 5α-Reductases and Bind to Androgen Receptor in Vitro. *Pharm. Pharmacol.* **2018**, *9*, 193–207. [CrossRef]
64. Iehlé, C.; Délos, S.; Guirou, O.; Tate, R.; Raynaud, J.P.; Martin, P.M. Human prostatic steroid 5 alpha-reductase isoforms–a comparative study of selective inhibitors. *J. Steroid Biochem. Mol. Biol.* **1995**, *54*, 273–279. [CrossRef]
65. Anderson, L. *European Union Herbal Monograph on Serenoa repens (W. Bartram) Small, Fructus*; Committee on Herbal Medicinal Products: London, UK, 2015.
66. Nahata, A.; Dixit, V.K. Evaluation of 5α-reductase inhibitory activity of certain herbs useful as antiandrogens. *Andrologia* **2014**, *46*, 592–601. [CrossRef]
67. Kang, X.C.; Chen, T.; Zhou, J.L.; Shen, P.Y.; Dai, S.H.; Gao, C.Q.; Zhang, J.Y.; Xiong, X.Y.; Liu, D.B. Phytosterols in hull-less pumpkin seed oil, rich in Δ7-phytosterols, ameliorate benign prostatic hyperplasia by lowing 5α-reductase and regulating balance between cell proliferation and apoptosis in rats. *Food Nutr. Res.* **2021**, *65*. [CrossRef] [PubMed]
68. Nattagh-Eshtivani, E.; Barghchi, H.; Pahlavani, N.; Barati, M.; Amiri, Y.; Fadel, A.; Khosravi, M.; Talebi, S.; Arzhang, P.; Ziaei, R.; et al. Biological and pharmacological effects and nutritional impact of phytosterols: A comprehensive review. *Phytother. Res.* **2022**, *36*, 299–322. [CrossRef]
69. Umetani, M.; Shaul, P. 27-Hydroxycholesterol: The first identified endogenous SERM. *Trends Endocrinol. Metab.* **2011**, *22*, 130–135. [CrossRef] [PubMed]
70. Nguyen, L.B.; Shefer, S.; Salen, G.; Tint, S.G.; Batta, A.K. Competitive inhibition of hepatic sterol 27-hydroxylase by sitosterol: Decreased activity in sitosterolemia. *Proc. Assoc. Am. Physicians* **1998**, *110*, 32–39.
71. Song, L.; Qu, D.; Zhang, Q.; Jiang, J.; Zhou, H.; Jiang, R.; Li, Y.; Zhang, Y.; Yan, H. Phytosterol esters attenuate hepatic steatosis in rats with non-alcoholic fatty liver disease rats fed a high-fat diet. *Sci. Rep.* **2017**, *7*, 41604. [CrossRef]
72. Kim, K.A.; Lee, I.A.; Gu, W.; Hyam, S.R.; Kim, D.H. β-Sitosterol attenuates high-fat diet-induced intestinal inflammation in mice by inhibiting the binding of lipopolysaccharide to toll-like receptor 4 in the NF-κB pathway. *Mol. Nutr. Food Res.* **2014**, *58*, 963–972. [CrossRef] [PubMed]
73. Liu, T.; Zhang, L.; Joo, D.; Sun, S.C. NF-κB signaling in inflammation. *Signal Transduct. Target Ther.* **2017**, *2*, 17023. [CrossRef]
74. Liao, P.C.; Lai, M.H.; Hsu, K.P.; Kuo, Y.H.; Chen, J.; Tsai, M.C.; Li, C.X.; Yin, X.J.; Jeyashoke, N.; Chao, L.K. Identification of β-Sitosterol as In Vitro Anti-Inflammatory Constituent in *Moringa oleifera*. *J. Agric. Food Chem.* **2018**, *66*, 10748–10759. [CrossRef]
75. Kasirzadeh, S.; Ghahremani, M.H.; Setayesh, N.; Jeivad, F.; Shadboorestan, A.; Taheri, A.; Beh-Pajooh, A.; Azadkhah Shalmani, A.; Ebadollahi-Natanzi, A.; Khan, A.; et al. β-Sitosterol Alters the Inflammatory Response in CLP Rat Model of Sepsis by Modulation of NF-κB Signaling. *Biomed. Res. Int.* **2021**, *2021*, 5535562. [CrossRef]
76. Yang, G.; An, H.J. β-sitosteryl-3-O-β-glucopyranoside isolated from the bark of *Sorbus commixta* ameliorates pro-inflammatory mediators in RAW 264.7 macrophages. *Immunopharmacol. Immunotoxicol.* **2014**, *36*, 70–77. [CrossRef] [PubMed]
77. Yuan, L.; Zhang, F.; Shen, M.; Jia, S.; Xie, J. Phytosterols Suppress Phagocytosis and Inhibit Inflammatory Mediators via ERK Pathway on LPS-Triggered Inflammatory Responses in RAW264.7 Macrophages and the Correlation with Their Structure. *Foods* **2019**, *8*, 582. [CrossRef] [PubMed]
78. Sudeep, H.V.; Venkatakrishna, K.; Amrutharaj, B.; Anitha; Shyamprasad, K. A phytosterol-enriched saw palmetto supercritical CO2 extract ameliorates testosterone-induced benign prostatic hyperplasia by regulating the inflammatory and apoptotic proteins in a rat model. *BMC Complement. Altern. Med.* **2019**, *19*, 270. [CrossRef]

79. Hamid, A.R.; Umbas, R.; Mochtar, C.A. Recent role of inflammation in prostate diseases: Chemoprevention development opportunity. *Acta. Med. Indones.* **2011**, *43*, 59–65.
80. Hussain, T.; Tan, B.; Yin, Y.; Blachier, F.; Tossou, M.C.; Rahu, N. Oxidative Stress and Inflammation: What Polyphenols Can Do for Us? *Oxid. Med. Cell. Longev.* **2016**, *2016*, 7432797. [CrossRef]
81. Ştefănescu, R.; Farczadi, L.; Huţanu, A.; Ősz, B.E.; Mărușteri, M.; Negroiu, A.; Vari, C.E. *Tribulus terrestris* Efficacy and Safety Concerns in Diabetes and Erectile Dysfunction, Assessed in an Experimental Model. *Plants* **2021**, *10*, 744. [CrossRef] [PubMed]
82. Salehi, B.; Quispe, C.; Sharifi-Rad, J.; Cruz-Martins, N.; Nigam, M.; Mishra, A.P.; Konovalov, D.A.; Orobinskaya, V.; Abu-Reidah, I.M.; Zam1, W.; et al. Phytosterols: From Preclinical Evidence to Potential Clinical Applications. *Front. Pharmacol.* **2021**, *11*, 599959. [CrossRef] [PubMed]
83. Van Rensburg, S.J.; Daniels, W.M.; van Zyl, J.M.; Taljaard, J.J. A comparative study of the effects of cholesterol, beta-sitosterol, beta-sitosterol glucoside, dehydroepiandrosterone sulphate and melatonin on in vitro lipid peroxidation. *Metab. Brain. Dis.* **2000**, *15*, 257–265. [CrossRef]
84. Chen, Z.; Wu, A.; Jin, H.; Liu, F. β-Sitosterol attenuates liver injury in a rat model of chronic alcohol intake. *Arch. Pharm. Res.* **2020**, *43*, 1197–1206. [CrossRef]
85. International, B.R. Retracted: Apoptosis and Molecular Targeting Therapy in Cancer. *Biomed. Res. Int.* **2020**, *2020*, 2451249. [CrossRef]
86. Megyesi, J.; Tarcsafalvi, A.; Seng, N.; Hodeify, R.; Price, P.M. Cdk2 phosphorylation of Bcl-xL after stress converts it to a pro-apoptotic protein mimicking Bax/Bak. *Cell Death Discov.* **2016**, *2*, 15066. [CrossRef]
87. Minutoli, L.; Rinaldi, M.; Marini, H.; Irrera, N.; Crea, G.; Lorenzini, C.; Puzzolo, D.; Valenti, A.; Pisani, A.; Adamo, E.B.; et al. Apoptotic Pathways Linked to Endocrine System as Potential Therapeutic Targets for Benign Prostatic Hyperplasia. *Int. J. Mol. Sci.* **2016**, *17*, 1311. [CrossRef]
88. Kyprianou, N.; Tu, H.; Jacobs, S.C. Apoptotic versus proliferative activities in human benign prostatic hyperplasia. *Hum. Pathol.* **1996**, *27*, 668–675. [CrossRef]
89. Rodríguez-Berriguete, G.; Fraile, B.; de Bethencourt, F.R.; Prieto-Folgado, A.; Bartolome, N.; Nuñez, C.; Prati, B.; Martínez-Onsurbe, P.; Olmedilla, G.; Paniagua, R.; et al. Role of IAPs in prostate cancer progression: Immunohistochemical study in normal and pathological (benign hyperplastic, prostatic intraepithelial neoplasia and cancer) human prostate. *BMC Cancer* **2010**, *10*, 18. [CrossRef]
90. Minutoli, L.; Altavilla, D.; Marini, H.; Rinaldi, M.; Irrera, N.; Pizzino, G.; Bitto, A.; Arena, S.; Cimino, S.; Squadrito, F.; et al. Inhibitors of apoptosis proteins in experimental benign prostatic hyperplasia: Effects of serenoa repens, selenium and lycopene. *J. Biomed. Sci.* **2014**, *21*, 19. [CrossRef] [PubMed]
91. Zhang, W.; Liu, H.T. MAPK signal pathways in the regulation of cell proliferation in mammalian cells. *Cell Res.* **2002**, *12*, 9–18. [CrossRef] [PubMed]
92. Ohtsuka, T.; Buchsbaum, D.; Oliver, P.; Makhija, S.; Kimberly, R.; Zhou, T. Synergistic induction of tumor cell apoptosis by death receptor antibody and chemotherapy agent through JNK/p38 and mitochondrial death pathway. *Oncogene* **2003**, *22*, 2034–2044. [CrossRef] [PubMed]
93. Davis, R.J. Signal transduction by the JNK group of MAP kinases. *Cell* **2000**, *103*, 239–252. [CrossRef]
94. Yue, J.; López, J.M. Understanding MAPK Signaling Pathways in Apoptosis. *Int. J. Mol. Sci.* **2020**, *21*, 2346. [CrossRef]
95. Papatsoris, A.G.; Papavassiliou, A.G. Molecular 'palpation' of BPH: A tale of MAPK signalling? *Trends Mol. Med.* **2001**, *7*, 288–292. [CrossRef]
96. Ropiquet, F.; Giri, D.; Lamb, D.J.; Ittmann, M. FGF7 and FGF2 are increased in benign prostatic hyperplasia and are associated with increased proliferation. *J. Urol.* **1999**, *162*, 595–599. [CrossRef]
97. Carson, C.; Rittmaster, R. The role of dihydrotestosterone in benign prostatic hyperplasia. *Urology* **2003**, *61* (Suppl. S1), 2–7. [CrossRef]
98. Wells, A.; Gupta, K.; Chang, P.; Swindle, S.; Glading, A.; Shiraha, H. Epidermal growth factor receptor-mediated motility in fibroblasts. *Microsc. Res. Tech.* **1998**, *43*, 395–411. [CrossRef]
99. Ahel, J.; Hudorović, N.; Vičić-Hudorović, V.; Nikles, H. Tgf-beta in the natural history of prostate cancer. *Acta Clin Croat.* **2019**, *58*, 128–138. [CrossRef] [PubMed]
100. Pugazhenthi, S.; Nesterova, A.; Sable, C.; Heidenreich, K.A.; Boxer, L.M.; Heasley, L.E.; Reusch, J.E. Akt/protein kinase B up-regulates Bcl-2 expression through cAMP-response element-binding protein. *J. Biol. Chem.* **2000**, *275*, 10761–10766. [CrossRef] [PubMed]
101. Von Holtz, R.L.; Fink, C.S.; Awad, A.B. beta-Sitosterol activates the sphingomyelin cycle and induces apoptosis in LNCaP human prostate cancer cells. *Nutr. Cancer.* **1998**, *32*, 8–12. [CrossRef] [PubMed]
102. Berges, R.R.; Windeler, J.; Trampisch, H.J.; Senge, T. Randomised, placebo-controlled, double-blind clinical trial of beta-sitosterol in patients with benign prostatic hyperplasia. Beta-sitosterol Study Group. *Lancet* **1995**, *345*, 1529–1532. [CrossRef] [PubMed]
103. Berges, R.R.; Kassen, A.; Senge, T. Treatment of symptomatic benign prostatic hyperplasia with beta-sitosterol: An 18-month follow-up. *BJU Int.* **2000**, *85*, 842–846. [CrossRef]
104. Klippel, K.F.; Hiltl, D.M.; Schipp, B. A multicentric, placebo-controlled, double-blind clinical trial of beta-sitosterol (phytosterol) for the treatment of benign prostatic hyperplasia. German BPH-Phyto Study group. *Br. J. Urol.* **1997**, *80*, 427–432. [CrossRef]

105. Wilt, T.; Ishani, A.; MacDonald, R.; Stark, G.; Mulrow, C.; Lau, J. Beta-sitosterols for benign prostatic hyperplasia. *Cochrane Database Syst. Rev.* **2000**, *1999*, CD001043. [CrossRef]
106. Coleman, C.I.; Hebert, J.H.; Reddy, P. The effect of phytosterols on quality of life in the treatment of benign prostatic hyperplasia. *Pharmacotherapy* **2002**, *22*, 1426–1432. [CrossRef]
107. Kadow, C.; Abrams, P.H. A double-blind trial of the effect of beta-sitosteryl glucoside (WA184) in the treatment of benign prostatic hyperplasia. *Eur. Urol.* **1986**, *12*, 187–189. [CrossRef]
108. Fischer, A.; Jurincic-Winkler, C.D.; Klippel, K.F. Conservative treatment of benign prostatic hyperplasia with high-dosage b-sitosterol (65 mg): Results of a placebo-controlled double-blind study. *Uroscopy* **1993**, *1*, 12–20.
109. Sudeep, H.V.; Thomas, J.V.; Shyamprasad, K. A double blind, placebo-controlled randomized comparative study on the efficacy of phytosterol-enriched and conventional saw palmetto oil in mitigating benign prostate hyperplasia and androgen deficiency. *BMC Urol.* **2020**, *20*, 86. [CrossRef] [PubMed]
110. Commission Regulation (EC) No 608/2004 of 31 March 2004. Concerning the Labelling of Foods and Food Ingredients with Added Phytosterols, Phytosterol Esters, Phytostanols and/or Phytostanol Esters (Text with EEA Relevance). 2004, 97. Available online: http://data.europa.eu/eli/reg/2004/608/oj/eng (accessed on 16 February 2023).
111. Ling, W.H.; Jones, P.J. Dietary phytosterols: A review of metabolism, benefits and side effects. *Life Sci.* **1995**, *57*, 195–206. [CrossRef]
112. Hendriks, H.F.; Brink, E.J.; Meijer, G.W.; Princen, H.M.; Ntanios, F.Y. Safety of long-term consumption of plant sterol esters-enriched spread. *Eur. J. Clin. Nutr.* **2003**, *57*, 681–692. [CrossRef] [PubMed]
113. Perini, M.; Paolini, M.; Camin, F.; Appendino, G.; Vitulo, F.; De Combarieu, E.; Sardone, N.; Martinelli, E.M.; Pace, R. Combined use of isotopic fingerprint and metabolomics analysis for the authentication of saw palmetto (*Serenoa repens*) extracts. *Fitoterapia* **2018**, *127*, 15–19. [CrossRef] [PubMed]
114. Saw Palmetto Extract | USP-NF. Available online: https://www.uspnf.com/official-text/revision-bulletins/saw-palmetto-extract (accessed on 16 February 2023).
115. European Pharmacopoeia 10.0 | PDF | Chemistry | Pharmaceutical. Available online: https://ro.scribd.com/document/508063535/European-Pharmacopoeia-10-0 (accessed on 16 February 2023).
116. Dreikorn, K. The role of phytotherapy in treating lower urinary tract symptoms and benign prostatic hyperplasia. *World J. Urol.* **2002**, *19*, 426–435. [CrossRef]
117. Giammarioli, S.; Boniglia, C.; Di Stasio, L.; Gargiulo, R.; Mosca, M.; Carratù, B. Phytosterols in supplements containing *Serenoa repens*: An example of variability of active principles in commercial plant-based products. *Nat. Prod. Res.* **2019**, *33*, 2257–2261. [CrossRef]
118. Kim, K.C.; Choi, M.-J.; Kim, K.-M.; Hill, W.S.; Dohnalek, M.; Davis, S.; Jung, J.-C. Analysis of *Serenoa repens* Extracts on Korean Market and Their Comparison to USP Saw Palmetto Extract Standards. *Food Suppl. Biomater. Health* **2021**, *1*, e26. [CrossRef]
119. Sorenson, W.R.; Sullivan, D. Determination of campesterol, stigmasterol, and beta-sitosterol in saw palmetto raw materials and dietary supplements by gas chromatography: Collaborative study. *J AOAC Int.* **2007**, *90*, 670–678. [CrossRef] [PubMed]
120. Penugonda, K.; Lindshield, B.L. Fatty Acid and Phytosterol Content of Commercial Saw Palmetto Supplements. *Nutrients* **2013**, *5*, 3617–3633. [CrossRef]
121. Sorenson, W.R.; Sullivan, D. Determination of campesterol, stigmasterol, and beta-sitosterol in saw palmetto raw materials and dietary supplements by gas chromatography: Single-laboratory validation. *J AOAC Int.* **2006**, *89*, 22–34. [CrossRef] [PubMed]
122. Fibigr, J.; Šatínský, D.; Solich, P. A UHPLC method for the rapid separation and quantification of phytosterols using tandem UV/Charged aerosol detection—A comparison of both detection techniques. *J. Pharm. Biomed. Anal.* **2017**, *140*, 274–280. [CrossRef] [PubMed]
123. Duong, S.; Strobel, N.; Buddhadasa, S.; Stockham, K.; Auldist, M.; Wales, B.; Orbell, J.; Cran, M. Rapid measurement of phytosterols in fortified food using gas chromatography with flame ionization detection. *Food Chem.* **2016**, *211*, 570–576. [CrossRef] [PubMed]
124. Du, M.; Ahn, D.U. Simultaneous Analysis of Tocopherols, Cholesterol, and Phytosterols Using Gas Chromatography. *J. Food Sci.* **2002**, *67*, 1696–1700. [CrossRef]
125. Vu, D.C.; Lei, Z.; Sumner, L.W.; Coggeshall, M.V.; Lin, C.-H. Identification and Quantification of Phytosterols in Black Walnut Kernels. *J. Food. Compost. Anal.* **2019**, *75*, 61–69. [CrossRef]
126. Zhang, T.; Wang, T.; Liu, R.; Chang, M.; Jin, Q.; Wang, X. Chemical Characterization of Fourteen Kinds of Novel Edible Oils: A Comparative Study Using Chemometrics. *LWT* **2020**, *118*, 108725. [CrossRef]
127. Alvarez-Sala, A.; Garcia-Llatas, G.; Cilla, A.; Barberá, R.; Sánchez-Siles, L.M.; Lagarda, M.J. Impact of Lipid Components and Emulsifiers on Plant Sterols Bioaccessibility from Milk-Based Fruit Beverages. *J. Agric. Food Chem.* **2016**, *64*, 5686–5691. [CrossRef]
128. Hamdan, I.J.A.; Claumarchirant, L.; Garcia-Llatas, G.; Alegría, A.; Lagarda, M.J. Sterols in infant formulas: Validation of a gas chromatographic method. *Int. J. Food Sci. Nutr.* **2017**, *68*, 695–703. [CrossRef]
129. Wang, M.; Zhang, L.; Wu, X.; Zhao, Y.; Wu, L.; Lu, B. Quantitative Determination of Free and Esterified Phytosterol Profile in Nuts and Seeds Commonly Consumed in China by SPE/GC–MS. *LWT* **2019**, *100*, 355–361. [CrossRef]
130. Claassen, F.W.; van de Haar, C.; van Beek, T.A.; Dorado, J.; Martinez-Inigo, M.J.; Sierra-Alvarez, R. Rapid Analysis of Apolar Low Molecular Weight Constituents in Wood Using High Pressure Liquid Chromatography with Evaporative Light Scattering Detection. *Phytochem. Anal.* **2000**, *11*, 251–256. [CrossRef]

131. Abidi, S.L. Chromatographic analysis of plant sterols in foods and vegetable oils. *J. Chromatogr. A* **2001**, *935*, 173–201. [CrossRef]
132. De Monte, C.; Carradori, S.; Granese, A.; Di Pierro, G.B.; Leonardo, C.; De Nunzio, C. Modern extraction techniques and their impact on the pharmacological profile of *Serenoa repens* extracts for the treatsment of lower urinary tract symptoms. *BMC Urol.* **2014**, *14*, 63. [CrossRef]
133. Islam, M.A.; Jeong, B.-G.; Kerr, W.L.; Chun, J. Validation of Phytosterol Analysis by Alkaline Hydrolysis and Trimethylsilyl Derivatization Coupled with Gas Chromatography for Rice Products. *J. Cereal Sci.* **2021**, *101*, 103305. [CrossRef]
134. Feng, S.; Wang, L.; Belwal, T.; Li, L.; Luo, Z. Phytosterols Extraction from Hickory (Carya Cathayensis Sarg.) Husk with a Green Direct Citric Acid Hydrolysis Extraction Method. *Food Chem.* **2020**, *315*, 126217. [CrossRef] [PubMed]
135. Feng, S.; Liu, S.; Luo, Z.; Tang, K. Direct saponification preparation and analysis of free and conjugated phytosterols in sugarcane (*Saccharum officinarum* L.) by reversed-phase high-performance liquid chromatography. *Food Chem.* **2015**, *181*, 9–14. [CrossRef]
136. Standard Methods for the Analysis of Oils, Fats and Derivatives—6th Edition. Available online: https://www.elsevier.com/books/standard-methods-for-the-analysis-of-oils-fats-and-derivatives/paquot/978-0-08-022379-7 (accessed on 17 February 2023).
137. Srigley, C.T.; Hansen, S.L.; Smith, S.A.; Abraham, A.; Bailey, E.; Chen, X.; Chooi, S.H.; Clement, L.M.; Dao, M.; Fardin Kia, A.R.; et al. Sterols and Stanols in Foods and Dietary Supplements Containing Added Phytosterols: A Collaborative Study. *J. Am. Oil Chem. Soc.* **2018**, *95*, 247–257. [CrossRef]
138. Phillips, K.M.; Ruggio, D.M.; Ashraf-Khorassani, M. Analysis of Steryl Glucosides in Foods and Dietary Supplements by Solid-Phase Extraction and Gas Chromatography. *J. Food Lipids* **2005**, *12*, 124–140. [CrossRef]
139. Sherma, J.; Rabel, F. Review of Advances in Planar Chromatography-Mass Spectrometry Published in the Period 2015–2019. *J. Liq. Chromatogr. Relat.* **2020**, *43*, 394–412. [CrossRef]
140. Lu, B.; Zhang, Y.; Wu, X.; Shi, J. Separation and determination of diversiform phytosterols in food materials using supercritical carbon dioxide extraction and ultraperformance liquid chromatography-atmospheric pressure chemical ionization-mass spectrometry. *Anal. Chim. Acta* **2007**, *588*, 50–63. [CrossRef] [PubMed]
141. Nzekoue, F.K.; Caprioli, G.; Ricciutelli, M.; Cortese, M.; Alesi, A.; Vittori, S.; Sagratini, G. Development of an Innovative Phytosterol Derivatization Method to Improve the HPLC-DAD Analysis and the ESI-MS Detection of Plant Sterols/Stanols. *Int. Food Res. J.* **2020**, *131*, 108998. [CrossRef]
142. Hryniewicka, M.; Starczewska, B.; Tkaczuk, N. Simple Approach Based on Ultrasound-Assisted Emulsification Microextraction for Determination of β-Sitosterol in Dietary Supplements and Selected Food Products. *Microchem. J.* **2020**, *155*, 104775. [CrossRef]
143. Shah, U.M.; Patel, S.M.; Patel, P.H.; Hingorani, L.; Jadhav, R.B. Development and Validation of a Simple Isocratic HPLC Method for Simultaneous Estimation of Phytosterols in Cissus quadrangularis. *Indian J. Pharm. Sci.* **2010**, *72*, 753–758. [CrossRef]
144. Özdemir, İ.S.; Dağ, Ç.; Özinanç, G.; Suçsoran, Ö.; Ertaş, E.; Bekiroğlu, S. Quantification of Sterols and Fatty Acids of Extra Virgin Olive Oils by FT-NIR Spectroscopy and Multivariate Statistical Analyses. *LWT* **2018**, *91*, 125–132. [CrossRef]
145. Liu, Y.; Wang, Y.; Xia, Z.; Wang, Y.; Wu, Y.; Gong, Z. Rapid Determination of Phytosterols by NIRS and Chemometric Methods. Spectrochim. Acta Part A Spectrochim. *Acta. A Mol. Biomol. Spectrosc.* **2019**, *211*, 336–341. [CrossRef] [PubMed]
146. Garcia-Llatas, G.; Alegría, A.; Barberá, R.; Cilla, A. Current Methodologies for Phytosterol Analysis in Foods. *Microchem. J.* **2021**, *168*, 106377. [CrossRef]
147. Gomes Silva, M.; Santos, V.S.; Fernandes, G.D.; Calligaris, G.A.; Santana, M.H.A.; Cardoso, L.P.; Ribeiro, A.P.B. Physical approach for a quantitative analysis of the phytosterols in free phytosterol-oil blends by X-ray Rietveld method. *Int. Food Res. J.* **2019**, *124*, 2–15. [CrossRef] [PubMed]
148. Nisca, A.; Ştefănescu, R.; Stegăruş, D.I.; Mare, A.D.; Farczadi, L.; Tanase, C. Phytochemical Profile and Biological Effects of Spruce (*Picea abies*) Bark Subjected to Ultrasound Assisted and Microwave-Assisted Extractions. *Plants* **2021**, *10*, 870. [CrossRef]
149. Lee, J.H.; Lee, K.; Jun, S.H.; Song, S.H.; Shin, C.H.; Song, J. A Multiplex Phytosterol Assay Utilizing Gas Chromatography-Mass Spectrometry for Diagnosis of Inherited Lipid Storage Disorders. *Ann. Lab. Med.* **2019**, *39*, 411–413. [CrossRef]
150. Almeling, S.; Ilko, D.; Holzgrabe, U. Charged Aerosol Detection in Pharmaceutical Analysis. *J. Pharm. Biomed. Anal.* **2012**, *69*, 50–63. [CrossRef]
151. Viinamäki, J.; Ojanperä, I. Photodiode Array to Charged Aerosol Detector Response Ratio Enables Comprehensive Quantitative Monitoring of Basic Drugs in Blood by Ultra-High Performance Liquid Chromatography. *Anal. Chim. Acta* **2015**, *865*, 1–7. [CrossRef]
152. Laakso, P. Analysis of Sterols from Various Food Matrices. *Eur. J. Lipid Sci. Technol.* **2005**, *107*, 402–410. [CrossRef]
153. Clement, L.M.; Hansen, S.L.; Costin, C.D.; Perri, G.L. Quantitation of Sterols and Steryl Esters in Fortified Foods and Beverages by GC/FID. *J. Am. Oil Chem. Soc.* **2010**, *87*, 973–980. [CrossRef]
154. Careri, M.; Elviri, L.; Mangia, A. Liquid Chromatography–UV Determination and Liquid Chromatography–Atmospheric Pressure Chemical Ionization Mass Spectrometric Characterization of Sitosterol and Stigmasterol in Soybean Oil. *J. Chromatogr. A* **2001**, *935*, 249–257. [CrossRef] [PubMed]
155. Maguire, L.S.; O'Sullivan, S.M.; Galvin, K.; O'Connor, T.P.; O'Brien, N.M. Fatty acid profile, tocopherol, squalene and phytosterol content of walnuts, almonds, peanuts, hazelnuts and the macadamia nut. *Int. J. Food Sci. Nutr.* **2004**, *55*, 171–178. [CrossRef]
156. Yuan, C.; Ju, Y.; Jin, R.; Ren, L.; Liu, X. Simultaneous HPLC–DAD Analysis of Tocopherols, Phytosterols, and Squalene in Vegetable Oil Deodorizer Distillates. *Chromatographia* **2015**, *78*, 273–278. [CrossRef]

157. Sedbare, R.; Raudone, L.; Zvikas, V.; Viskelis, J.; Liaudanskas, M.; Janulis, V. Development and Validation of the UPLC-DAD Methodology for the Detection of Triterpenoids and Phytosterols in Fruit Samples of *Vaccinium macrocarpon* Aiton and *Vaccinium oxycoccos* L. *Molecules* **2022**, *27*, 4403. [CrossRef]
158. Acworth, I.N.; Bailey, B.; Plante, M.; Gamache, P. Simple and Direct Analysis of Phytosterols in Red Palm Oil by Reverse Phase HPLC and Charged Aerosol Detection. *Planta Med.* **2011**, *77*, PA2. [CrossRef]
159. Nair, V.D.; Kanfer, I.; Hoogmartens, J. Determination of stigmasterol, beta-sitosterol and stigmastanol in oral dosage forms using high performance liquid chromatography with evaporative light scattering detection. *J. Pharm. Biomed. Anal.* **2006**, *41*, 731–737. [CrossRef]
160. Slavin, M.; Yu, L. A Single Extraction and HPLC Procedure for Simultaneous Analysis of Phytosterols, Tocopherols and Lutein in Soybeans. *Food Chem.* **2012**, *135*, 2789–2795. [CrossRef]
161. Ito, M.; Ishimaru, M.; Shibata, T.; Hatate, H.; Tanaka, R. High-Performance Liquid Chromatography with Fluorescence Detection for Simultaneous Analysis of Phytosterols (Stigmasterol, β-Sitosterol, Campesterol, Ergosterol, and Fucosterol) and Cholesterol in Plant Foods. *Food Anal. Methods* **2017**, *10*, 2692–2699. [CrossRef]
162. Horník, S.; Sajfrtová, M.; Karban, J.; Sýkora, J.; Březinová, A.; Wimmer, Z. LC-NMR Technique in the Analysis of Phytosterols in Natural Extracts. *J. Anal. Methods Chem.* **2013**, *2013*, 526818. [CrossRef] [PubMed]
163. Gachumi, G.; El-Aneed, A. Mass Spectrometric Approaches for the Analysis of Phytosterols in Biological Samples. *J. Agric. Food Chem.* **2017**, *65*, 10141–10156. [CrossRef]
164. Cañabate-Díaz, B.; Segura Carretero, A.; Fernández-Gutiérrez, A.; Belmonte Vega, A.; Garrido Frenich, A.; Martínez Vidal, J.L.; Duran Martos, J. Separation and Determination of Sterols in Olive Oil by HPLC-MS. *Food Chem.* **2007**, *102*, 593–598. [CrossRef]
165. Baila-Rueda, L.; Cenarro, A.; Cofán, M.; Orera, I.; Barcelo-Batllori, S.; Pocoví, M.; Ros, E.; Civeira, F.; Nerín, C.; Domeño, C. Simultaneous Determination of Oxysterols, Phytosterols and Cholesterol Precursors by High Performance Liquid Chromatography Tandem Mass Spectrometry in Human Serum. *Anal. Methods* **2013**, *5*, 2249–2257. [CrossRef]
166. Sánchez-Machado, D.I.; López-Hernández, J.; Paseiro-Losada, P.; López-Cervantes, J. An HPLC method for the quantification of sterols in edible seaweeds. *Biomed. Chromatogr.* **2004**, *18*, 183–190. [CrossRef] [PubMed]
167. Peters, F.T.; Remane, D. Aspects of Matrix Effects in Applications of Liquid Chromatography–Mass Spectrometry to Forensic and Clinical Toxicology—A Review. *Anal. Bioanal. Chem.* **2012**, *403*, 2155–2172. [CrossRef]

MDPI AG
Grosspeteranlage 5
4052 Basel
Switzerland
Tel.: +41 61 683 77 34

Plants Editorial Office
E-mail: plants@mdpi.com
www.mdpi.com/journal/plants

www.ingramcontent.com/pod-product-compliance
Lightning Source LLC
Chambersburg PA
CBHW060408220926
51850CB00003B/6
9783725846894